Theory and Practice of Soil Loss Control in Eastern China

J.C. Zhang · D.L. DeAngelis · J.Y. Zhuang

Theory and Practice of Soil Loss Control in Eastern China

Foreword by Wen-Yue Hsiung

Springer

J.C. Zhang
Dean of the College of Forest Resources and Environment
Nanjing Forestry University
Nanjing, China
nfujczhang@sina.com

D.L. DeAngelis
Department of Biology
University of Miami
Coral Gables, FL, USA
ddeangelis@bio.miami.edu

J.Y. Zhuang
College of Forest Resources and Environment
Nanjing Forestry University
Nanjing, China
zjiayao@msn.com

ISBN 978-1-4419-9678-7 e-ISBN 978-1-4419-9679-4
DOI 10.1007/978-1-4419-9679-4
Springer New York Dordrecht Heidelberg London

Library of Congress Control Number: 2011928152

Printed on acid-free paper

Springer is part of Springer Science+Business Media (www.springer.com)

Foreword

Historically, dense sub-tropical, evergreen, broadleaf forest and mixed forest of conifer and broadleaf trees covered the middle and lower reaches of the Yangtze River basin. This area has the largest lakes in China, which, because they can store flood water, form a harmonious system with the Yangtze River. But when humans began to have impacts on the hills through time, the forests were seriously damaged. The natural forest was replaced by plantation forest for wood utilization, which was usually of low forest quality and low biodiversity, as well as suffering from serious water and soil losses, and causing frequent flooding. This situation seriously affected the sustainable development of society, both economically and in terms of regional ecological safety. In addition to regional physical and hydrological conditions that cause flood hazards, the combination of land reclamation and population growth created an environmental bomb, which resulted in the worst flooding in China since 1954 in the 1998 Yangtze River flood. The flood claimed more than 2000 lives and engendered economic losses of US $25,000 million. The 1998 Yangtze deluge was measured to be a 150-year flood, but the water flux was only a 60-year event. Scientists were inclined to attribute this extreme event to El Niño, a periodic warming of the eastern Pacific that began in 1997 and extended through the first half of 1998. Nevertheless, both observational and theoretical studies have proven that the destruction of natural vegetation cover, such as destructive lumbering of forests and over-cultivation and overgrazing of grassland, has been one of the major causes for the deterioration of the regional climate and environment.

Before 1980 in China, forest reforestation implied cultivating plantation forest for wood and for economics with few tree types. After the 1980s, the Chinese forestry system consisted of a developed forestry industry, and maintaining a healthy ecological system was determined as the main aim and guiding idea for a sustainable forestry development strategy. China is an agricultural country with 700 million farmers, most of whom were poor and used wood for construction, and used grass, shrubs, twigs, and litter of trees and straw as fuel for cooking. There was a strong desire to get rid of poverty from these mountainous areas and other areas of limited arable land. Based on local conditions, it was urgent for peasants, technical workers, and officials to reforest bare land, hilly, or highly sloping land that had formerly been cultivated, and land suffering desertification due to serious water and soil loss in southern China. It should be pointed out that the subsidies from the government for reforestation of steep hilly land that was formerly cultivated land are just enough for 3–5 years' income of the same cultivated land, and reforestation on bare land only provides money to farmers for weeding and management. Reforestation and economic interaction work should improve not only the ecological environment but also the living standards of local residents, which in turn will increase the awareness of the concept of environment protection of the local residents. Finally interaction among reforestation, community, and the economy can get rid of vicious cycle of the poverty trap.

This book *Theory and Practice of Soil Loss Control in Eastern China* by my good friend and coworker Prof. Zhang Jinchi, Prof. Donald DeAngelis, and Dr Zhuang Jiayao concerns not only the development of models for soil loss prediction but also reforestation techniques for water and soil loss control in the hilly and mountainous areas in the lower reaches of the Yangtze River, which have been carried out creatively and with great effect as a multi-functioning system by forestry scientific workers. With the development of society and economics of the Yangtze River delta, demands from residents for a good ecological environment have greatly increased. Thus, synergistic interaction between mountainous and hilly regions in the high-elevation catchments, such as the Dabie Mountains and the Daba Mountains, which can produce a good environment, keep water clean and protect the lower reaches from flooding – but lack money – and urban areas in the lower reaches, such as Nanjing, Wuxi, and Shanghai, which need clean water from the mountains urgently, but do know how to get it effectively, can be reached. This book also proposes that interactions between urban areas in the lower reaches, which provide money and technology, and the countryside, where ecological restoration for water and soil loss control with the use of reforestation, biogas, and solar and water energy, should be formed in the lower reaches of the Yangtze River in eastern China. This will decrease the differences in wealth between the countryside and the city and achieve an economic integration of city–countryside. In this sense, this book is good not only for technical workers in soil erosion control and forestry but also for strategies of managers in all the undeveloped countries of the world.

Nanjing, China Wen-Yue Hsiung

Preface

Not only is much of the region of the middle and lower reaches of the Yangtze River characterized by steep hills but also it contains the densest population and most economically developed area in China. Due to intense impact of human activities, forest resources have suffered serious damage, resulting in a low quality of forest, made up of a few types of plantation forest with simple structure and low capacity for water and soil loss control. In addition, due to the steep gradients of the hills, high susceptibility indices of soil erodibility, and plentiful rainfall of non-uniform distribution, intense water and soil loss, as well as landslides, occurs frequently. These geomorphologic and natural environmental characteristics determine that for the currently bare lands, steep hillslopes currently under cultivation, and the lands that have undergone desertification, reforestation is the only choice for achieving regional sustainable development and the realization of a harmonious relationship between man and nature.

Up until recently, research on forestry ecological engineering for water and soil loss control has made great progress on small areas or on single forest communities, but there is still a lack of synthetic research and model demonstration at the catchment scale. Because of the long timescales in forest development, the theory and methods of reforestation were mostly reached by means of substituting space for time in basic research, so it was technically difficult to make comparisons across space. More recently, researchers have performed blocks of experiments and demonstrations of reforestation in the provinces of Jiangxi, Anhui, Jiangsu, Zhejiang, etc., and acquired much support from Chinese national research projects over a period of 20 years and achieved much success. Based on the past research results and monitoring data over this long period, this book first presents the several proposed soil loss prediction models.

For soil loss control in the mountains of China, the first step is the determination of appropriate strategies. There is an urgent need for soil loss models that can provide sufficient information for making strategies to control soil loss using field observation data taken over short periods, such as several years. The USLE (universal soil loss equation) models were developed in the USA in the 1960s and applied for average annual soil loss prediction widely in the world. But their application requires observational data over many years. In addition, the USLE models do not take much account of the typical operations of human beings that are usual in China's cultivated lands. With the development of economy, controlling the environmental problems has become so important that more information should be provided for making soil loss control strategies based on field observation data over just a few years. Furthermore, soil loss control needs the cooperation of farmers and local residents. Effective soil loss models facilitate the communication between environmental experts and farmers and citizens.

This research also proposed a systematic application of the GOIUG (GIS-based observed instantaneous unit graph) model, integrated to make graphical predictions of instantaneous suspended sediment discharge with GIS, the IUG method, and a hydrologic model, which is used to simulate the suspended sediment generation and its transmission to the outlet. It helps to elucidate, quantitatively, the source of sediment discharge. Furthermore, a model of ER (effective rainfall erosivity)-USLE was developed from USLE to predict annual soil loss based on single events with an effective rain erosivity factor. The model portrays the interactions among seasonal precipitation, seasonal crop coverage, and individual operations by human beings (P_S). The litter factor is incorporated into USLE in order to create a FUSLE model for application in forests. These models are believed applicable with higher accuracy in China and suitable for strategy making in soil loss control in the cultivated land and forest management.

Second, this work developed many types of reforestation, especially some key techniques, such as secondary forest culture with a focused tree plantation method, agriculture–forestry methods in the hills, reforestation in extremely eroded areas of red soil, and forest soil fertility protection. These methods are providing technical support for nationally important projects, such as the Chinese natural forest protection, reforestation in cultivated land at slopes above 25°, and wood forestation forest construction in southern China.

In particular, techniques of reforestation also focused on improving the living standards of farmers, proposing a "small recycling" system made up of tree, crop–feeding–biogas–fertility–fishing, etc and a "big recycling" system made up of interaction between the countryside of mountainous areas and the urban areas in the lower elevations. The "small recycling" can not only make money for the farmers effectively, while keeping the environment clean from being polluted by animal excrement, but also save fuel materials such as twigs and litter of tree, and shrub and grass from being collected, thus protecting the reforested young trees from harm. The "big recycling" makes a harmonious society, with the aim of common health and wealth.

We have a long cooperation in eco-restoration with world famous ecologist, Professor Donald DeAngelis. During his recent visit in China, he was very pleased with the rapidly increasing forest coverage in south China. He thought the fast reforestation method may not only be useful in China but also provide a good model to eliminate poverty for the poor farmers in the other undeveloped countries all over the world. In addition, effective reforestation will enhance the effort to slow world climate warming. Then we discussed the possibility of compiling a book on reforestation techniques and soil loss control theory for the purpose of providing useful assistance to researchers, university graduates, and foresters. Both of us believed that is was the right time to compile a book integrating our common research progress and assessment on reforestation and soil loss control in eastern China.

Nanjing, China J.C. Zhang

Acknowledgments

I would like to thank Prof. Du Tianzheng of Jiangxi Agricultural University, Prof. Yu Mukui of Chinese forestry academy, Prof. Fang Yanming of Nanjing Forestry University, Mr Cheng Pen and Mr Fu Jun, Mr Zhao Xueshi, and Mr Xu Jianmin of Forestry Department of Anhui province for their support. I would like to express my gratitude to Prof. Hiroyuki Nakamura, Prof. Yoshiharu Ishikawa, and assistant Prof. Katsushige Shiraki of Tokyo University of Agriculture and Technology for providing valuable insights in developing new soil loss models for effective soil loss prediction. And I appreciate those forest and soil loss technicians as well as the farmers who conducted most of the field work described in this book. Special acknowledgements are also given to the sponsors and providers of funds for their support for the research, resulting in the publication of this book. My project team thanks the Natural Science Foundation of China (Project Nos 30872072 and 30872076), the financial support from the Construction Project of Excellent Subject for universities in Jiangsu province, China, the Chinese National 11th Five-Year Plan of Forestry Science (Project No. 2006BAD03A16), Foundation of Chinese Forest Ecosystem Services-Technology for Observation and Evaluation (Project No. 200704005/wb02-03), and the Nanjing Forestry University for their financial support of our research.

Contents

Part I
General Characteristics of the Hilly Region of Middle and Lower Yangtze River

Chapter 1
Ecological and Environmental Characteristics in the Hilly Region of Middle and Lower Yangtze River

Abstract The Yangtze River is one of the most important rivers in the world. Its middle and lower reaches lie in the subtropical monsoon area of east Asia, which has a warm and moist climate with clear distinction between the four seasons. The vegetation in the middle and lower reaches of Yangtze River takes on conspicuous vertical zoning characteristics. From low to high elevation, the vegetation makes a transition from evergreen broadleaf forest to mixed evergreen broadleaf and deciduous forest. The species diversity of forest plantations in the middle and lower reaches of the Yangtze River has very important status in China. The present problems of soil erosion are based mainly on the following factors: (1) plentiful precipitation provides strong force for water erosion; (2) the inhomogeneities in the temporal and spatial distributions of precipitation result in both frequent flood and drought calamities; (3) simple forest structure and monocultures of trees cause fragility of the forest ecosystem; (4) the sharp increase of population has made the forest destruction more serious and has caused flooding. Because water and soil loss has constrained the development of eastern China, the control of soil loss has been deemed as of primary importance in reforestation.

1.1 Introduction

Land use plays an important role in the phenomena associated with global change (Zhang et al., 1999). It is directly related to food security, urbanization, biodiversity, trans-boundary migration, environmental refugees, water and soil quality, runoff and sedimentation rates (Turner, 1989). Over thousands of years, land use and land use change have been transforming the ecosystems of Yangtze valley, which is now the home of 400 million people.

The Yangtze River is one of the most important rivers in the world. It is the third longest and ninth largest in basin area (Fig. 1.1). Its water discharge is the largest in the western Pacific Ocean and the fifth largest in the world, and its sediment load is the fourth largest in the world (Zhang and Hu, 1996).

The Yangtze River originates in the Qinghai-Tibet Plateau and extends about 6300 km eastward to the northern East China Sea at around 31°30′N and 121°30′E. The headwaters of the trunk stream are located 5100 m above the sea level, which implies a mean longitudinal profile gradient of 0.08%. It flows through 11 provinces: Qinghai, Tibet, Sichuan, Yunnan, Chongqing, Hubei, Hunan, Jiangxi, Anhui, Jiangsu, and Shanghai, in that order. The drainage basin is located between

J.C. Zhang et al., *Theory and Practice of Soil Loss Control in Eastern China*,
DOI 10.1007/978-1-4419-9679-4_1, © Springer Science+Business Media, LLC 2011

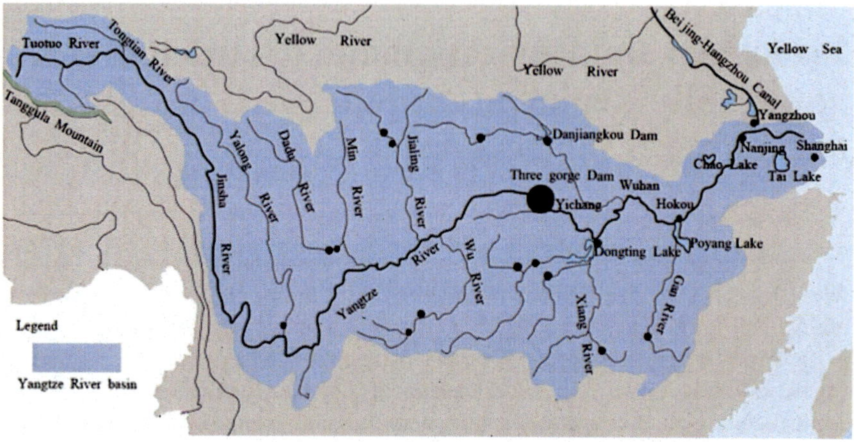

Fig. 1.1 The Yangtze watershed and its tributaries

24.5 and 35.5°N and the climate is characteristic of the subtropical and warm–wet zones. The mean precipitation is 1000–1400 mm/year and the evaporative power (maximum potential evaporation) is 700–800 mm/year in the drainage area. Precipitation and runoff vary with season, with 70–80% being distributed in the rainy season from May to October. Its catchment basin covers an area of 1,808,500 km^2 (18.8% of the territory of China) and has a population of more than 400 million (one-third of the Chinese population). This chapter concerns the ecological and environmental characteristics of the hilly region of the middle and lower reaches of the Yangtze River.

The middle and lower reaches of the Yangtze River consist of the southern region of Huaihe River, the Daba Mountains, the eastern region of the Wuling Mountains, and the northern region of the Nanling Mountains (Fig. 1.2). Regarding the administrative range, it includes most parts of Hubei, Hunan and Jiangxi provinces, etc., and parts of Anhui, Zhejiang, and Jiangsu provinces, as well as Shanghai (Fig. 1.3). The total land area of this region is 91.03 km $\times$ 10^4 km, accounting for 9.5% of China's whole territory. Because of the dense population, large agricultural population (31.3% of the total agricultural population in China), and relatively small arable area (only 18.8% of China's total arable area), most of the low mountains and hills with relatively better water and soil resources have been reclaimed to dry cropland, resulting in a high cultivation index. Granite, gneiss, purple shale, and limestone are common parent rocks in this region. Soil conditions vary considerably here. On the one hand, on the vast hilly region to the south of the Yangtze River is distributed mainly red and yellow soil, which is clayey, acidic, and thin. Soil organic matter does not readily accumulate, and thus the soil is low in yield. On the other hand, the region north of the Yangtze River has mainly yellow brown and yellow cinnamon soil with poor structure, high clayey texture, and weak ability for water storage and drought resistance.

Fig. 1.2 The main rivers and mountains in China

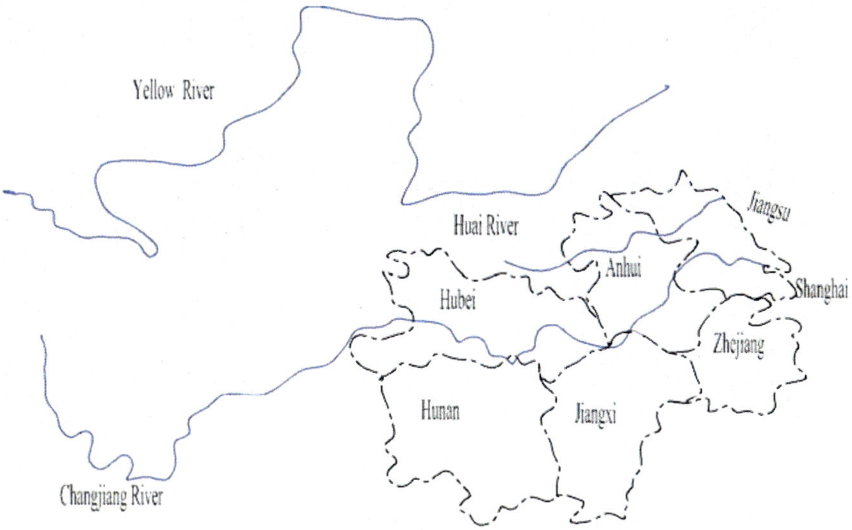

Fig. 1.3 The administrative divisions in east China

1.2 Physiographic Conditions

Being a subtropical monsoon area of east Asia, this region has a warm and moist climate with clear distinction between the four seasons; it is hot in summer and cold in winter, with a long free-frost period. Both precipitation and temperature

Table 1.1 Areas of mountains, hills, and plains in each province in the middle and lower reaches of Yangtze River

Provinces	The proportion each geomorphic type takes up in total land area (%)		
	Mountain	Hill	Plain
Anhui Province	31.2	37.5	39.3
Jiangxi Province	30.4	44.4	25.2
Jiangsu Province	–	15.0	85.0
Hunan Province	51.2	29.3	19.5
Zhejiang Province	21.3	49.1	29.6
Hubei Province	55.5	24.5	20.0

are rather high here with an annual average precipitation of 1100 mm and up to 1800–2000 mm in some areas. It has two rainy seasons: one is the plum rain season between spring and summer and the other is the typhoon rain season in summer. This area ranks as one of the regions that have the longest rainy seasons in China, which is very beneficial for the growth of crops and forests. As a result, the growing season in this region is long.

The terrain of China is higher in the west and lower in the east, which is like a three-step ladder. The middle and lower reaches of the Yangtze River consist mainly of plains, hills, and mountains, with low mountains and hills accounting for three-fourths of the whole area. The areas beside its trunk channel and tributaries have low elevations, making the third terrain of the three large terrains in China. The area around Yidu City along the Yangtze River consists mainly of plains, with the watershed as its periphery, low mountains and hills of Huaiyang in the north, and low mountains and hills of Jiangnan in the south (Zhou, 2005).

Hills and mountains cover more than 65% of the total territory of most provinces, except in Jiangsu Province and Shanghai (Table 1.1). Larger cordilleras run mostly from northeast to southwest. These include the Wuling cordillera in the west of Hunan Province, the Xufeng cordillera, the Luoxiao cordillera on the border of Hunan and Jiangxi provinces, the Mufu cordillera on the border of Jiangxi, Hubei, and Hunan provinces, the Wuyi cordillera on the border of Fujian and Jiangxi provinces, the Huayu cordillera near Zhejiang and Jiangxi, and others. The cordilleras that run from northwest to southeast are the Daba cordillera between Hubei Province and Chongqing City, and the Dabie Mountain cordillera between Hubei and Anhui provinces. The Nanling Mountains run across Hunan, Jiangxi, and Guangdong provinces from east to west. Between these cordilleras and plains there are widely distributed hills, including the Jiangnan hills in the middle and southern part of Hunan and Jiangxi provinces, the western part of Zhejiang Province, and the southern part of Anhui Province, covering an area of 37 km × 10^4 km.

The plains in the middle and lower reaches of Yangtze River include the Jianghan plain, Dongting Lake plain, Poyang Lake plain, the plain in eastern Hubei Province along the Yangtze River, the plain and delta in Jiangsu and Anhui provinces along the Yangtze River, and some lakes and coastal plain. The Jianghan plain is made up of Yidu, Jingzhou, Shishou, Jiayu, Wuhan, Tianmen, Zhongxiang City, etc, with an elevation range of 20–40 m. Because the ground level of the river bank is 3–15 m lower than the historical floodwater level, flood calamities occur frequently and the flood control situation is very serious. The Dongting Lake plain consists of the U-shaped delta and alluvial plain, which is enclosed by Dongting Lake, Jinshi, Yiyang, Miluo, and Yueyang, and is open to the north. The elevation of this plain is 22–32 m, 20 m lower than the historical flood level, with the 2,625-km^2 Dongting Lake playing an important role in adjusting and delaying floods. The plain along the Yangtze River in eastern Hubei Province is made up of high valleys with flat and first grade terrace topography. With a general elevation of 12–25 m, it has intermittent dikes constructed along the Yangtze River. The Poyang Lake plain includes the Poyang Lake and its surrounding alluvial plain. The elevation along the lake is 14–17 m, generally 6–10 m lower than flood level. With an area of 3913 km^2, it can regulate and store water during the flooding season and provide supplementary water for the lower reaches of Yangtze River during the low water season – it is a typical lake, which both receives intake water and discharges water. The plain in Jiangsu and Anhui provinces along the Yangtze River includes the alluvial plain along the Yangtze River between the cities of Jiujiang and Nanjing and its tributaries. Its elevation is 8–20 m, 1–8 m lower than the flood level. There are many lakes on both banks of the Yangtze River, among which the largest is Chaohu Lake on the northern bank, with an area of 780 km^2. The delta and littoral plain consists of northern and southern Jiangsu Province, northern Zhejiang Province, and Shanghai below Zhenjiang City. Generally speaking, its elevation is 2–5 m, 0.5–3 m lower than flood level. There are relatively more lakes on the southern bank, with Tai Lake being the largest, with an area of 2460 km^2 (Zhou, 2005).

The area in the middle and lower reaches of the Yangtze River crosses five tectonic units – the Qinling fold system, the Yangzi Para platform, the south China fold system, the Sino-Korean Para platform, and the southeast coastal fold system – and belongs to the Paleozoic structural belt and plateau cover fold belt in the phase of Yanshan.

With the advantage of such environmental characteristics as the geomorphology of low mountains and hills, of soil, of climate resources, and so forth, the region in middle and lower reaches of the Yangtze River is not only an area of concentration of subtropical evergreen broadleaf forests but also a base for the cultivation of timber and pulp forests, as part of the system of plantation forests in China. Warm and moist natural conditions nurture rich and varied natural vegetation and a complicated forest ecosystem in this region. The virgin forest in the south of this region is evergreen broadleaf forest, while vegetation in the north displays transition characteristics from the subtropical zone to the temperate zone. Vegetation biodiversity has a highly important status in China.

1.3 The Vegetation Characteristics

1.3.1 Distribution of Vegetation in the Middle and Lower Reaches of the Yangtze River

The virgin vegetation in the south of this region is evergreen broadleaf forest, the major components of which include *Castanopsis*, *Lithocarpus*, and *Cyclobalanopsis* in the beech family; *Cinnamomum*, *Phoebe*, and *Machilus* in the laurel family; *Schima* and *Camellia* in the tea family; and *Manglietia* and *Michelia* in the Magnolia family. In addition, *Ilex* and *Symplocos* are also commonly found there. In Anhui Province, there are 3320 kinds of angiosperms, 13.3% of the total 25,000 kinds in China; 72 kinds of gymnosperms, 36% of the total 200 kinds in China; 253 kinds of ferns, 9.7% of the total 2600 kinds in China; 600 kinds of bryophytes, 27.3% of the total 2200 kinds in China. There are 6455 kinds of major plants cultivated in Anhui Province, such as the mung bean in Mingguang, hickory nut of Ningguo, chrysanthemums in Huizhou, and papaya in Xuanzhou. All of them are unique varieties, famous both at home and abroad. In recent years, the area of artificial forest in the middle and lower reaches of Yangtze River has been expanding steadily. Influenced by the plantation forests, vegetation has been restored quickly. Taking Taihe County of Jiangxi Province as an example, before the year 1991 when the plantation forests planting program was started, there were only seven plant species, most of which were light-loving and drought-resistant graminaceous plants, such as green bristlegrass herb (*Setaria viridis*), Arundinella hirta (*Arundinella hirta*), Cherokee rose (*Rosa laevigata*) and Smilax china (*Smilax china*). After plantation forestation, the number of species increased to 21 in 1993 and 58 in 2001. Some foreign plants, which originated in tropical America and North America, such as licorice (*Glycyrrhiza uralensis*) and annual fleabane herb (*Erigeron annuus*), are beginning to enter the plantation forests.

The vegetation in the middle and lower reaches of Yangtze River takes on conspicuous vertical zoning characteristics. From low elevation to high location, the vegetation makes a transition from evergreen broadleaf forest to mixed evergreen broadleaf and deciduous forest (Ke et al., 2002) (Fig. 1.4). The conifer forests in this region are distributed as an "inlay" type, that is, interweaving into other forest types or mixing with the broadleaf trees. Conifer forests here are mostly thermophilic ones, among which Chinese fir (*Cunninghamia lanceolata*) and Masson pine (*Pinus massoniana*) are the main species grown here. In addition, the Mao bamboo (*Phyllostachys edulis*) forest occupies a certain area in this region in a natural mosaic pattern. The vegetation in the north has the characteristics of transition from subtropical zone to temperate zone. On the one hand, the broadleaf forest includes deciduous forests, which are the most common, and evergreen ones, which are mainly distributed south of the Yangtze River. Of the deciduous forest types, Liaotungensis genera (*Quercus* L.), Chinese beech (*Fagus* L.), poplar (*Populus* L.), etc. are the most popular for plantation forests. Because of the differences in micrometeorology between hills and valleys, the evergreen broadleaf forest consists of different varieties and proportions. The major evergreen

Fig. 1.4 Distribution of vegetation types in the Yangtze River watershed. *I*, Broadleaf evergreen vegetation; *II*, broadleaf deciduous vegetation; *III*, needle-leaved evergreen vegetation; *IV*, dwarf shrubby; *V*, high-cold steppes and meadows; *VI*, alpine sparse vegetation; *VII*, cultivated vegetation; *VIII*, no data area

broadleaf trees are Cyclobalanopsis oak (*Cyclobalanopsis glauca*), bitter mesophanerophytes (*Castanopsis sclerophylla*), rock oak (*Lithocarpus glaber*), Purple phoebe (*Machilus sheareri*), and red phoebe (*Machilus thunbergii*), etc. The coniferous trees include both warmth-loving ones such as masson pine, Chinese fir, Bashan Torreya (*Torreya fargesii*), and cedarwood (*Cupressus funebri*), and cold-resistant ones, such as spruce (*Picea brachytyla*).

1.3.2 The Characteristics of Biodiversity in the Middle and Lower Reaches of the Yangtze River

The species diversity of forest vegetation in the middle and lower reaches of the Yangtze River has very important status in China. Statistics shows that in China there are 337 families of seed plants, 3200 genera, and 26,276–27,268 species (Li, 1996). There are 52 families of fern, 204 genera, and 2,600 species (according to the statistics of the editorial committee of "Chinese Physical Geography" of Chinese Academy of Sciences). There are 2457 species of bryophytes (Redfearn et al., 1996). The number of species of seed plants in China is the third in the world, which is next to Malaysia (about 45,000 species) and Brazil (about 40,000 species). The flora of Hubei and Hunan provinces, in the middle and lower Yangtze River, is the characteristic of the central China district, and that in Zhejiang and Anhui provinces is the characteristic of the east China district. According to statistics (Qi, 1993; Liu, 1995), there are 207 families of seed plants, 1279 genera, and 6370 species in central China, and there are 174 families of seed plant, 1180 genera, and 4259 species in east China. At the family level, central China accounts for almost two-thirds of the flora of China as a whole, and east China accounts for more than one-half of

Table 1.2 Bryophyte diversity in Hunan, Jiangxi, Hubei, and Anhui provinces

Province	Liverwort (family/genus/species)	Moss (family/genus/species)	Total (family/genus/species)	Documents
Hunan	24/34/70	39/110/197	63/144/269	Rao et al. (1997)
Jiangxi	26/43/89	40/109/216	66/152/305	Fang et al. (1998)
Hubei	–	/ /272	–	Redfearn et al. (1996)
Anhui	–	/ /457	–	Redfearn et al. (1996)

the flora of China as a whole. At the genus level, the flora of the four provinces accounts for 40% in central China and 37% in east China. Obviously, the diversity of biological pedigree is quite high. Table 1.2 shows that the biodiversity in the four provinces is considerable. According to the above analysis, the estimated total amount of vegetation, including seed plants, ferns, and bryophytes, is 10,000 species. High biodiversity is the foundation for complex structure and diverse functions within forests, as well as a resource bank of valuable species for the forestry industry.

1.3.3 Current Situation and Existing Problems of Biodiversity

Despite the fact that the species diversity of forest plants in the four provinces is adequate and there are abundant types of plants, the actual situation is not encouraging. For various reasons, many of the existing biodiversity conservation areas are constrained. A large number of species are confined to the limited natural conservation areas, extinction of some species is unavoidable, and the management measure of biodiversity has lagged behind. Present problems are focused mainly on the following aspects:

(1) There is abundant plantation forest, but wild forest area is sparse, with simplified and fragile ecosystems.

In the past 10 years, the forest coverage rate has increased by a large amount (Ke et al., 2003) (Fig. 1.5), through the forestation projects called "Afforestation of the Barren Land Mountains" and "Shelter Forest Engineering Construction Along the Yangtze River." For example, the forest coverage of Jiangxi Province rose from 40.3% in 1991 to 50.8% in 1994, and the forest coverage of Hunan Province in 1997 was up to 51.41%, and it rose to 25.97% in 1994 in Hubei Province.

The first stage of the project of "Shelter Forest Engineering Construction Along the Yangtze River" involves 27 counties in Jiangxi Province. By 1997, about 1,216,000 ha bare lands were afforested, of which about 119,000 ha were in Xingguo County. However, from the point of view of ecosystem diversity, the species of trees and the structure of forest are not reasonable in this region. The area of the natural forest is limited, and the majority of new cultivated forest is plantation forest. With respect to forest structure, the proportion of material forest is high,

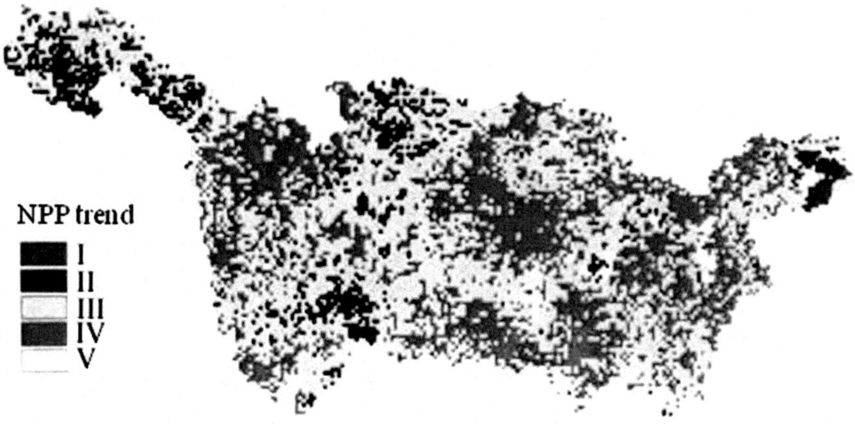

Fig. 1.5 Variation of vegetation in the Yangtze watershed from the year 1980 to 2000. *I*, Decreased significantly; *II*, decreased, though not significantly; *III*, increased, though not significantly; *IV*, increased significantly; *V*, no data area

which accounts for 60–70%, while the proportion of shelter forest is low, accounting for 5–9%. With respect to tree species composition, the proportion of the conifer trees is high, accounting for 74–79%, while the proportion of the broadleaf forest is low, accounting for 21–26%. Undoubtedly, large areas of homogeneous forest with simplified tree species will result in simplified ecosystems. In the past 10 years, due to the campaign for natural environment conservation, which is advocated actively by Chinese government, natural reserves and forest parks have increased remarkably, but this area is still small. There are 24 natural reserves in Hubei Province, the area of which is 220,000 ha, accounting for 1.78% of the territory area. Extant natural forest, especially evergreen broadleaf forest, is surrounded and fragmented by plantation ecosystems suffering from severe anthropogenic disturbances. Therefore, its structure is fragile, with limited integral ecological function. The vulnerability of the forest ecosystem is clear.

(2) The area of broadleaf forest has decreased, and some species have nearly disappeared. According to the forest resource investigation of Jiangxi Province, the area of broadleaf forest decreased from 1,679,000 ha in 1955 to 1,149,000 ha in 1994, of which the arboreal broadleaf forest was mostly secondary forest or plantation forest afforested after the felling of natural forest. A lot of species have disappeared, because the suitable habitat was lost as the broadleaf forest was destroyed, leading to the partial disappearance and extinction of species that had low adaptability. Fortunately, the trend of forest disappearance is being controlled effectively because the area of forest is now increasing.

(3) The distributions of species and plant genetic diversity are decreasing. Because of forest destruction and the fragmentation of the ecosystem structure, many species that previously had a continuous range have became disconnected. Because of

the destruction of habitat diversity, many species have become confined to single habitats, whereas they could in principle be distributed over many habitats. On special sites, anthropogenic disturbances, such as excavating Chinese herbal medicine and selective felling of trees, have caused species originally at high population densities to become thinner and species originally at low population density to become rarer. Red Phoebe, Min Phoebe (*Phoebe bournei*) and Camphorwood (*Cinnamomum camphora)* are typical representatives of evergreen broadleaf forest in the middle subtropical zone, which were distributed along low-altitude valleys and rivers and formed stable communities with other plant types. In the low-altitude areas, the anthropogenic activities are so intense that such typical communities have already become rare. Schima superb (*Schima superba*) and sweet mesophanerophytes (*Castanopsis eyrei*) have high-density populations in the evergreen broadleaf forest but have become rare in many areas.

(4) The fragmentation of forest into patchy landscapes has caused geographic isolation of communities. Because of extensive cultivation and development, many evergreen communities of broadleaf forest have already disappeared in the low hill area. Big orchards, non-irrigated farmlands, and economic forests have taken their place. Some patches of secondary evergreen broadleaf forest are distributed in the boundary areas of catchments of larger mountain systems, with little dispersal between them. Regarding the horizontal pattern, the communities have already effectively become islands, which cannot be organically connected, and finally have become geographically isolated. The biodiversity is menaced at the levels of genetic, community, ecosystem, and landscape diversity.

The disconnected habitat, destroyed habitat, and unreasonable artificial harvesting have decreased the genetic diversity or even led to extinctions. Those consequences have been caused by non-natural activity, which can be unfavorable to human beings and which violates the rule that society should develop in harmony with nature. Genetic diversity of plant communities is still not well understood, but the present research indicates that the level of genetic diversity is low.

(5) There are potentially abundant biodiversity resources, but there have been management and utilization modes tending toward single species. The abundant biodiversity resources in middle and lower reaches of the Yangtze River are the potential foundations of forestry, as well as a foundation of maintaining the region's sustainable development. The most serious problems involve the single afforested species, that is, monocultures of small plants that can be utilized.

1.3.4 Biodiversity Crisis

Many unforeseen crises may arise with the destruction of biodiversity. The hierarchical structure of biodiversity at the genetic, population, and ecosystem levels is the result of biological evolution over billions of years, which is the natural legacy.

Biodiversity has the external value of offering environmental services of physical material (food, fuel, timber, fiber, medicine, etc.), as well as genetic information and spiritual values, etc. But the destruction of biodiversity and the extinction of species destroy the knowledge stored in the DNA of living cells. With respect to utilizable resources, biodiversity includes natural resources that are the basis of economic development, which can improve the people's material life. Therefore, destruction of biodiversity is a waste of natural resources. From the viewpoint of ecology, nature is a complicated system, which includes different interrelated processes and components. Destruction of biodiversity makes the structure of ecosystem fragile and decreases its function of resisting natural disasters.

1.4 Soil Characteristics and Nutrient Status

1.4.1 Main Soil Types

Because of the long-term influence of geologic activity, the physiognomy and landforms of the middle and lower Yangtze River are very complicated. The main physiognomy consists of mountains, hills, plateaus, and basins. The area of plains is small, and the low mountains and hills account for more than 70% of the whole area. The elevations of mountains in the region of Anhui, Zhengjiang, Jiangxi, and Hunan provinces are between 1500 and 2000 m, and the mountains are formed of granite, rhyolite, sandstone, and limestone. The hill region includes areas in southern Anhui Province, northern Zhejiang Province, the valley and hill area of the Ganjiang River, the hill area surrounding Dongting and Poyang lakes, and basins and hills in the middle of Hunan Province, where the parent materials are sandstone, shale, and Quaternary Period laterite as the core. The plains are alluvial in the middle and lower Yangtze River, as are the plains surrounding the lakes, where the parent materials are alluvium sediment, etc.

There are three main types of soil in the hilly region of the middle and lower Yangtze River: the red and yellow soil of the Ferralsol soil order, yellow brown soil of Alfisol soil order, and purple and lime soil of Lithosol soil order (Xiong and Li, 1987; Li, 1983).

The red earth is distributed in the wide low mountains and hilly region south of the Yangtze River, including the majority of Jiangxi and Hunan provinces, northern Fujian Province, southern Anhui, Jiangsu, Zhejiang provinces, and part of the region of Hubei Province. Of the above hilly regions, the hilly regions of Dongting and Poyang Lakes are the most centralized, where the parent materials are Quaternary Period laterite, sandstone, granite, etc. The red soil is formed under the climatic conditions of the subtropical zone, that is, under a warm climate of adequate precipitation and a long frost-free period. The natural vegetation of the red soil is evergreen broadleaf forest in the subtropical zone. According to the conditions of formation, the process of formation, the fertility, and the characteristics of how the red soil can be utilized, it can be divided into four subordinate kinds of red earth:

lateritic red soil, dark red earth, yellow red earth, and brown red soil. The yellow soil is distributed in such areas as Hubei, Hunan, Jiangxi, Fujian, Zhejiang, and Anhui provinces. The parent materials are granite, sandstone, gneiss, quartzite, shale, limestone, and sediments of the Tertiary and Quaternary Periods, etc.

1.4.2 The Soil Characteristics and Existing Problems

1.4.2.1 Soil Organic Matter Does not Readily Accumulate, and the Soil Has Obvious Characteristics of Acidity and Leanness

The vast areas of the hilly region are the main areas on which red and yellow soils occur south of the Yangtze River, and where the soil has low productivity characteristics, such as acidity, and does not readily accumulate organic matter, etc., which slow down the processes of vegetation restoration and reconstruction.

The red soil has the outstanding function of high aluminum accumulation such that the ratio of calcium to aluminum is 1.8–2.2%, and the main clay mineral is kaolinite. The organic content in topsoil is generally between 1 and 1.5%, the total nitrogen varies from 0.09 to 0.12%, and the total phosphorus is about 0.06%, while the content of soluble reactive phosphorus is low. Its base value is highly unsaturated, and cation exchange capacity (CEC) varies from 0.6 to 4.5 cmol (+)/kg. The content of active aluminum in the soil is high, and the value of pH is from 4.5 to 5.2. The content of humus in the top layer of red soil under forest cover is high, and the thickness of the soil layer varies considerably (Xiong and Li, 1987; Li, 1983).

Due to the characteristics of the parent rock, the purple soil has such characteristics as strong physical weathering, weak chemical weathering, and sustained eluviation of calcium carbonate. Purple soil sections are mostly mauve, mauve brown, and purple dark brown, the whole section of which has the same color without conspicuous differences. A part of purple soil contains calcium carbonate, whose content can be up to 10% and the pH value of which is between 7.5 and 8.5. In most of the purple soil, the content of calcium carbonate in topsoil layer is lower than 1%. Because its parent rock has good water permeability or its soil has been formed for long time, most calcium carbonate has been eluviated. In purple soil, the organic content is commonly lower than 1%, although some areas are exceptions to this rule. The content of phosphorus and potassium in purple soil is high. Its cation exchange capacity is high, and its degree of base saturation is above 80 and 90%.

Lime soil is formed under the influence of carbonate rock. Its formation process and characteristics have a close relation with eluviation and weathering of carbonate rock. Under moist climatic conditions, the carbonate is eluviated consistently in the formation process of lime soil, while the weathering of rock can supply calcium consistently. Eluviation and accumulation of calcium frequently occur together. The black lime soil is either in an initial stage of calcium loss, or the restoration of calcium has been relatively strong. Red lime soil has strong losses of calcium and silicon and accumulation of aluminum. Brown lime soil is somewhere between the above two types. The lime soil layer is shallow, and the boundary between rock and soil is obvious when the slope is steep. When the vegetation has been destroyed or utilized improperly, it is easily eroded. In lime soil, the organic matter in soil decomposes rapidly and the soil structure worsens quickly with inadequate aboveground water. Thus, its utilization for agriculture and forestation is difficult.

1.4.2.2 Forest Vegetation Has Been Damaged, and Its Function of Water Infiltration and Storage Is Weak

The forest litter has an important function of storing water, which can hold a large quantity of runoff. The capability of forest litter for holding water shows slight variation among different types of forests (Table 1.3).

Usually litter's average water-holding capacity is about three times its weight. For broadleaf forests, the water-holding capacity of the litter is usually higher than three times of its own weight, and for coniferous trees and Mao bamboo, it is usually under three times. The average capability of holding water is 36.62 t ha^{-1}, and the greatest holding water capacity for one precipitation event can be up to 0.60–8.06 mm, while the average value is 3.69 mm (Liu, 1996).

The ability to hold water by soil in forests is determined by the soil structure and porosity. In mixed evergreen and deciduous broadleaf forest, broadleaf and conifer forest, and evergreen broadleaf forest, the condition of soil porosity is better than that in the monoculture forest because branches and leaves in the former forest types are luxuriant, with much litter, and the root system is well developed, which facilitates the development of soil porosity. Once the forest vegetation is destroyed, the physical characteristics of the soil, especially porosity and soil structure, will worsen, which has a substantial influence on the hydrological ecological effect (Table 1.4). The ability of soil to hold or store water is closely related to total soil porosity (Table 1.5).

Table 1.3 Capability of litter for holding water for different types of forests

	Holding water rate (%)		Holding water capacity			
			t hm^{-2}		mm	
Forest type	Range	Average	Range	Average	Range	Average
Mixed evergreen and deciduous broadleaf forest	366.6–461.2	413.92	45.20–80.62	62.92	4.52–8.06	6.29
Mixed conifer and broadleaf forest	291.6–331.2	307.08	22.98–76.44	42.96	2.30–7.64	4.30
Deciduous broadleaf forest	290.3–366.6	338.09	25.37–42.13	35.16	2.54–4.21	3.52
Evergreen broadleaf forest	318.3–457.7	388.65	23.27–54.73	33.28	2.33–5.47	3.33
Masson pine forest	201.0–318.7	306.66	12.67–51.20	32.18	1.27–5.12	3.22
P. edulis forest	103.0–331.2	207.19	6.00–36.55	16.29	0.60–3.66	1.62
Chinese fir forest	176.2–335.1	261.98	3.95–25.97	13.92	0.40–4.09	1.64

Table 1.4 Soil porosity in different types of forests

	Non-capillary pore (%)		Capillary pore (%)		Total pore (%)	
Forest type	Range	Average	Range	Average	Range	Average
Mixed evergreen and deciduous broadleaf forest	21.3–26.1	23.7	49.5–60.5	55.0	75.6–81.7	78.7
Mixed conifer and broadleaf forest	14.0–24.0	20.6	40.7–49.1	43.9	54.7–73.0	64.6
Deciduous broadleaf forest	16.8–19.7	18.2	43.8–61.4	52.1	61.9–81.1	70.3
Evergreen broadleaf forest	9.7–21.5	13.3	37.5–56.7	46.3	47.2–72.1	59.6
Masson pine forest	8.2–24.9	15.5	35.5–61.5	44.4	48.5–81.1	59.8
P. edulis forest	7.9–10.7	9.0	40.3–50.5	45.4	51.0–58.7	54.4
Chinese fir forest	11.7–20.0	16.3	33.8–49.3	43.3	50.8–63.5	59.6
Grassland	5.2–7.5	6.4	24.8–40.1	32.4	30.1–47.5	38.8

Table 1.5 Water-holding capacity in different types of forests

		Holding water capacity			
		Maximum		Minimum	
Forest and soil type	Bulk density (g cm^{-3})	%	mm	%	mm
Evergreen broadleaf forest – red soil in the mountain region	0.74	112.5	499.7	76.6	340.0
Mixed evergreen and deciduous broadleaf forest – yellow soil	0.61	123.9	453.4	42.8	156.7
Deciduous broadleaf forest – yellow soil	0.71	101.5	432.3	30.2	128.8
Mixed conifer and broadleaf forest – yellow soil	0.92	71.5	394.6	26.1	144.2
Plantation forest of *Pinus taeda*– red soil	1.18.	50.6	358.7	37.0	262.0
Masson pine forest – red soil	1.15	51.0	352.2	7.2	49.4
P. edulis forest – red soil	1.24	41.8	310.9	32.9	224.9

Because non-capillary pores are the main pathways through which the soil gravity water moves, the water-storing capability of the soil is closely related to non-capillary pores. The water-holding ability is usually above 100 mm for evergreen broadleaf forest soil and mixed evergreen and deciduous broadleaf forest soil in subtropical areas. Water-holding capacity for forest ecosystems depends on non-capillary porosity, which accounts for above 90% of all porosity. The second component of water-holding capacity is that of forest litter, whose capacity ranges from 3 to 10 mm. The ability of soil to redistribute water is maximized only under the condition of maintaining a good forest and forest litter.

1.4.2.3 The Content of Soil Nutrients Is Low, and the Limitation on Forest Growth Is Strong

The content of soil nutrient in the hilly region of the middle and lower Yangtze River is relatively low, especially in the conifer forest. The content of soil organic matter is between 15.50 and 72.87 g kg^{-1}, and the contents of total nitrogen and soluble reactive phosphorus are low. The lack of soluble reactive phosphorus in soil may be limiting factors for forest growing (Table 1.6).

Compared with the average values of total quantity of Fe in soils in China (3%), the values of total quantity of Fe in this area is several times higher. The values of total quantities of Mo and Cu in soil in this region are also high, amounting to 46.7 and 4.194 mg kg^{-1}, respectively. The average values of total quantities of B, Zn, and Mn are low, amounting to 38.6, 84.7 and 274 mg kg^{-1}, respectively. The effective Fe, which is 41.7 mg kg^{-1} in the soil, is much higher than the threshold value of 3 mg kg^{-1} of the normal agro-chemical standard. The average value of available Cu, which is 0.97 mg kg^{-1} in this area, is smaller than the threshold value of 2 mg kg^{-1}. The content of available Zn, which is 2.58 mg kg^{-1}, is larger than its threshold value of 2.00 mg kg^{-1}. The content of available Mn is moderate. The values of available B and Mo, which are 0.30 and 0.10 mg kg^{-1}, respectively, are much lower than their critical values of 0.50 and 0.15 mg kg^{-1}, respectively. Lack of soil nutrients influences vegetation restoration and forest growth.

Table 1.6 Concentrations of soil nutrients

Area	Forest type	Organic matter (g kg^{-1})	Total nitrogen (g kg^{-1})	Quick-acting phosphorus (mg kg^{-1})	Quick-acting potassium (mg kg^{-1})
	Secondary *Quercus variabilis* forest of 25a	27.42	3.70	10.41	121.84
Xiashu of Jiangsu Province	Chinese fir forest of 25a	37.21	1.79	6.88	93.08
	P. taeda forest of 25a	38.46	1.72	7.34	62.28
	P. edulis forest	30.98	1.56	4.32	40.15
	P. taeda forest of 20a	42.10	2.30	2.18	237.0
	Masson pine forest of 20a	67.70	3.40	5.72	372.4
Jingdezhen of Jiangxi Province	Natural broadleaf forest of 20a	81.10	4.30	7.04	191.0
	P. taeda forest of 3a	15.50	1.10	0.84	141.5
	P. edulis forest of 15a	41.30	1.70	1.76	111.6
	Chinese fir forest	35.38	1.78	0.75	83.56
	Masson pine forest	38.90	1.96	0.94	85.12
Jian'ou of Fujian Province	Castanopsis forest	72.87	3.64	1.27	98.45
	Mixed forest of Chinese fir and *S. superba*	47.91	2.37	1.06	89.56

1.4.2.4 Forest Vegetation Is Destroyed, the Soil Nutrient Is Eluviated, and Soil Fertility Is Seriously Degraded in Artificial Forest

The red soil and yellow soil regions in the middle and lower Yangtze River have been suffering from serious soil erosion that is second only to that of the Loess Plateau area. Because of unreasonable utilization of forests, the soil erosion has aggravated. The area of soil erosion in Hunan Province rose from 18,700 km^2 in the 1950s to 47,200 km^2 in the 1990s, accounting for 21.18% of the territory area in the whole province, whose annual soil loss was 1.8×10^8 t. In the 1950s, 1960s, and 1970s, the soil erosion area accounted for 6, 10, and 12.9% of the whole area of Jiangxi Province, respectively. But in the 1980s, it accounted for 20.7%. By the middle period of the 1990s, it increased sharply to 21.1%, with the soil erosion area amounting to 35,200 km^2. The area of soil erosion in Anhui Province rose from 20,600 km^2 in the 1970s to 33,600 km^2 in 1987, accounting for 24% of the whole territory area. The direct consequence of soil erosion led to fertility and soil losses in the top layers, and a large amount of nutrient flowing into the river and reservoirs. The soil became thin, and the amount of nutrient was reduced, while forest productivity dropped. In Zhushan County in Hubei Province, because of the cultivation of wasteland with steep slopes, the loss of soil nutrient matter was equivalent to 14,000 t of $(NH_4)_2SO_4$ every year; it was 1.47 times the volume of annual fertilizer sales. In this area, natural broadleaf forest was reduced and plantation forest increased and was distributed with a high average density, especially for short-cycle plantation forest for industry. Production activity, such as machinery felling and transporting, made the soils to harden and physical condition to worsen. Studies showed that the content of organic matter, soluble reactive phosphorus, and soluble reactive potassium in topsoil in Chinese fir forest and loblolly pine (*P. taeda*) forest is lower than that in the forest of oak tree (*Quercus L.*) (Luo and Gao, 1994). If Chinese fir trees were afforested in the same place continuously, they could make the land fertility drop such that the chemical indices of the soil would drop by 10–20% and by 40–50% in the second and third generation, respectively (Yu, 1998; Yu, 2000).

1.5 The Current Situation of Soil Erosion and Its Causes

1.5.1 The Current Situation of Soil Erosion

Historically, there were many evergreen broadleaf forests and mixed forests of conifer and broadleaf trees in the subtropical zone in the middle and lower Yangtze River, and soil erosion was slight. But after the Han Dynasty, people began to cut forests for crop cultivation when the population increased. Virgin forest was destroyed, and soil erosion aggravated gradually. According to investigations, the area of soil erosion has increased on average by 1.25% every year in 13 counties in the Yangtze River valley. The soil erosion range in Hunan Province expanded from 34 counties in the 1950s to 87 counties in 1995, rising from 18,700 to 47,200 km^2, amounting to 21.18% of total territory area of the whole province with annual soil loss of 180 million tons. The area of cultivated land suffering from wind erosion and water erosion is 3147 km^2. The area of soil erosion in the 1980s was 3.2 times the area during the 1950s. Currently the average annual soil erosion is 4393 t km^{-2} yr^{-1}, and the area of moderate soil erosion accounts for 65% of the whole area of soil erosion. The area of soil erosion in Hubei Province rose to 67,600 km^2 in 1989 from 50,120 km^2 in 1957, which extends over all 58 counties in the hilly region. The area of moderate soil erosion to more intense erosion accounts for 58% of the total area of erosion. The area of soil erosion in Anhui Province rose from 20,631 km^2 in 1970s to 33,600 km^2 in 1987, increasing by 62.9% in 17 years, accounting for 24% of the area of the whole territory, among which the area of soil erosion in the Dabie Mountains reached 7449.3 km^2.

For the Xiangjiang, Ruan, Zi, and Li rivers in the midstream of the Yangtze River, the quantity of transported sand increased by 10% during the last 10 years, during which the sediment concentrations were 0.179, 0.129, 0.229, and 0.416 kg m^{-3}, respectively, for the four rivers. The quantity of sand transported into Dongting Lake increased by 34.4% from the 1950s to the 1980s. The average quantity of transported sand in the Yangtze River at the Yichang station was 5.2×10^8 t $year^{-1}$ from 1950s to 1979, while the average sediment concentration was 1.17 kg m^{-3}. The average quantity of transported sediment increased to 8.37×10^8 t $year^{-1}$, while the average sediment concentration was 1.39 kg m^{-3} from 1980 to 1985, among which the highest average sediment concentration was 1.88 kg m^{-3} in 1981. In the Poyang Lake valley, the sediment concentration in the Gan, Fu, Rao, Xiu, and Xinjiang rivers increased by 12.6, 24.5, 42.6, 88.9, and 2.7%, respectively, from the 1950s to the 1980s.

1.5.2 Factors Causing Soil Erosion

1.5.2.1 Plentiful Precipitation Provides Strong Force for Water Erosion

The middle and lower Yangtze River watershed is in the subtropical monsoon climate region of eastern Asia, where it is hot in summer and cold in winter, while the frost-free period is long, making a clear distinction between four seasons. In this area, temperature and precipitation are adequate and the growing season is long. It is one of the regions with the longest annual period of precipitation in China. Especially in summer, the warm–wet air coming from the ocean controls the climate of this basin, and produces seasonal precipitation, which usually causes big floods. Strong precipitation increases the force of soil erosion. According to investigation carried out in Jiangxi, Anhui, Jiangsu, and Zhejiang provinces, the surface runoff and soil losses in barren land were 3–6 times and 782 times those of forested land, respectively, while the soil losses during the rainy season accounted for 70–80%.

The annual average precipitation in the middle and lower Yangtze River is between 1000 and 2000 mm, which has two peak periods, one is the plum rain

Table 1.7 The precipitation in each province in the middle and lower Yangtze River

Province	Average annual precipitation (mm)	Max. daily precipitation (mm)	Rainy area
Anhui Province	700–1700	287.7	The landform rain in low mountains and hilly region
Jiangxi Province	1350–2130	327.4	Plum rain season in the end of spring and rainstorm in summer
Jiangsu Province	800–1200	314.3	Plum rain season in the end of spring and the typhoon rain in late summer
Hunan Province	1250–2090	259.5	Rainstorms at the beginning of summer account for 40% of whole year's precipitation
Zhejiang Province	850–1700	253.7	Plum rain season from the beginning of June to the beginning of July and typhoon rain from the end of August to the end of September
Hubei Province	750–1500	317.4	Period from June to August, accounts for 35–50% of whole year's precipitation
Shanghai	1000	204.4	Plum rain season in the end of spring and the typhoon rain in late summer

season at the end of spring and the other is the typhoon rain period in summer. During the rainy season, rainstorms are frequent, and the largest daily precipitation can reach 300 mm (Table 1.7). The distribution of precipitation is influenced largely by high mountains and local landforms. So the spatio-temporal distribution of soil erosion by water in eastern China varies with season and geomorphology.

1.5.2.2 Temporal and Spatial Distribution of Precipitation: The Inhomogeneous Nature of the Rainfall Causes Frequent Flood and Drought Calamities

The middle and lower Yangtze River has a subtropical monsoon climate, where warm, wet air currents form precipitation. Annual precipitation in this zone is 1000–1600 mm, which decreases gradually from southeast to northwest. Besides the influence of topography and monsoon climate, the plains, rivers, lakes, and mountains divide the area into several districts (Table 1.8).

The extensive tall mountain systems stop the warm, wet air currents and form abundant precipitation. For example, the Shennongjia, Dabie, and Mufu mountains usually have high precipitation, where the annual average precipitation is between 1600 and 2000 mm. In the area of Poyang Lake and the Ganjiang River valley, the annual average precipitation is 1400 mm, while the precipitation can reach 1000 mm in the rainy season. The average annual precipitation is between 1250 and 1450 mm in the Xiangjiang River valley and Dongting Lake, respectively.

Table 1.8 Distribution of precipitation in middle and lower Yangtze River

Region	Average annual precipitation	Rainfall in rainy season (April – October)	Rainfall in drought season (November – March)
Shennongjia area, Dajia Mountain, Yellow Mountain, Tianmu Mountain, Wuyi Mountain, Xuefeng Mountain, Wuling Mountain, etc.	1600–1200	1200–1600	300–400
Ganjing River valley	1400–1550	1050–1200	370–470
Poyang Lake region	1400–1600	1000–1150	350–450
Xiang River valley	1250–1450	900–1000	300–400
Dongting Lake region	1250–1360	950–1000	300–400
Both banks of the middle and lower reaches of the Yangtze River	1000–1350	800–1000	250–380

1.5.2.3 Simple Forest Structure and Monocultures of Trees Cause Fragility of the Forest Ecosystem

The terrain of middle and lower Yangtze River is relatively low in elevation, and agriculture is well developed. There are large areas of cultivated land with high multiple crop indices in the plains areas. Cultivated land accounts for 34.1% of the whole territory area in Anhui Province, and large areas of land in the hilly region have become cultivated lands and orchards.

Although the forestland in each province in the middle and lower reaches of the Yangtze River is increasing gradually, there are still some outstanding problems, which are as follows:

(1) The ratio of different forest types is not reasonable. The proportion of material (used for timber, pulp, etc.) forest is too high, while shelter forest is insufficient. For example, the proportion of shelter forest is 8% in Jiangxi Province and 6.3% in Hunan Province. Insufficient shelter forest has weakened the protective ability of forests and resulted in deterioration of the environment and in frequent calamities.
(2) The composition of species of trees is not reasonable. The coniferous trees of Chinese fir and Masson pine are absolutely predominant in the current forest, while the area of broadleaf forest and mixed forest has dropped sharply.
(3) The age structure of the forest is not reasonable. In the material-producing forest, the proportion of forest with young and middle age is high, while mature forest resources are insufficient. For example, the proportion of forest of young and middle age to that of mature age is 15.2 and 11.3 times, respectively, showing that the forests in young and middle age are absolutely predominant.

In the forest that is young in age, the amount of shade is low, the litter layer is thin, and the water-conserving function of soil is weak.

(4) The management mode is unreasonable. Since the 1980s, local government regarded planting artificial forest plantations and orchards as a good method to shake off poverty of local residents in the mountainous area in middle and lower Yangtze River, so plantation forests and orchards expanded rapidly. Farmers used the traditional management method, such as cultivation and grass clearing, which caused serious water and soil erosion. According to field observations, annual soil loss in shrub forest was 4 t ha^{-1} and it was 25 t ha^{-1} in orchards on steep land.

(5) There are many forests of low value. For a long time, the forest resource was destroyed repeatedly, which has formed much low-value forest. Fertility of forestland decreased and the ecological value as well as economic value became very low.

1.5.2.4 The Sharp Increase of Population Has Made the Forest Destruction More Serious and Has Caused Flooding

The middle and lower Yangtze River is one of the areas that have the densest population in China, accounting for 31.3% of national agricultural population with only 18.8% of the country's cultivated land. With such a population, the human pressure on the land is intensive.

To begin with, the huge water storage area in the Yangtze River basin has been largely reduced by irrational land use. As the tributaries which form the Yangtze River emerge from the mountainous regions into the middle Yangtze plain, a complex system of lakes and wetlands serves as a natural catchment area – one of the main natural defenses against flooding. But over the past four decades, irrational land use has weakened this system. The storage capacity of the Dongting Lake, the main catchment of the middle Yangtze, has shrunk from 29.3 billion m^3 in 1949 to 17.8 billion m^3 in recent times (He, 1998). In the 1950s, there were 1066 freshwater lakes in the area around the middle and lower Yangtze plains. But by the early 1990s, the number had shrunk to only 182, and the water area had been slashed by 46%; of 86,000 reservoirs in the middle and lower Yangtze regions, 40% had been silted up.

Second, the Yangtze River basin has lost 85% of its original forest cover (Zhang et al., 1999). Owing to increasing population, industrialization, and urbanization, the natural ecosystems across the Yangtze basin, such as forest, grassland, and wetlands, have been encroached upon by farmland and other man-made ecosystems on a large scale.

Especially in Sichuan Province, forest cover has decreased from 19 to 6% in 1949. The forests that once absorbed and held huge quantities of monsoon rainfall, which could then percolate slowly into the ground, are now largely gone. Forest cover clear-cutting resulted in heavy runoff from torrential rains and the continuous transport of massive quantities of silt into the river, which raises the water level and which, in turn, forces flood control officials to constantly raise the level of the surrounding levees.

Additionally the construction of factories, houses, and roads in the basin is increasing at a staggering pace. Land hunger forces more and more homes to be built on the river flood plain.

In addition to regional physical and hydrological conditions that cause flood hazards (e.g., Yangtze River's lower elevations and its four big branches at the confluence of the Yangtze River), the combination of land reclamation and population growth created an environmental bomb, which resulted in the Yangtze River's worst flood since 1954 in 1998. The flood claimed more than 2000 lives and engendered economic losses of US $25,000 million (Du, 1998).

The news media reported the 1998 Yangtze deluge as a 150-year flood, but the flux was only a 60-year event (He, 1998). Scientists were inclined to attribute this extreme event to El Niño, a periodic warming of the eastern Pacific that began in 1997 and extended through the first half of 1998 (Abu-Zeid, 1998). Nevertheless, both observational and theoretical studies have proved that the destruction of natural vegetation cover, such as destructive lumbering of forests and over-cultivation and overgrazing of grassland, has been one of the major causes for the deterioration of the regional climate and environment (Belt, 1975; Lean and Warrilow, 1989; Nobre et al., 1991; Pielke et al., 1991).

Statistics showed that the population in the area of Tai Lake rose from 19,300,000 in 1930 to 31,440,000 in 1984. In the past 10 years, the increase in population decreased the average area of cultivated land per person from 800.4 m^2 at the beginning of the 1980s to less than 667 m^2. The hilly region turned into a focal area of development, the results of which people should pay close attention to. The unreasonable cultivation and utilization of land caused the forest vegetation to suffer from serious destruction. There are 4,980,000 ha of cultivated land in Hubei Province, among which slopes between 15° and 25° account for 600,000 ha, or 12.05%, of the cultivated land. The slopes of 500,000 ha of cultivated land are above 25°, accounting for 10.4% of the total cultivated land. The forestland was 9,346,000 ha in 1949 in Jiangxi Province, and the forest coverage was 56.0%. During the Chinese "Campaign to make steel" in 1958, much wood was used as fuel. The campaign to learn from Dazhai in agriculture in the 1970s emphasized the importance of food production from the mountains and hills, in which shrub, marsh, and much forestlands were turned into cultivated land in the mountains and hills. In the 1980s, the government permitted the farmers to manage forestry by family unit, just as in agriculture, and this caused a new disaster of felling trees as material for construction. These unsustainable measures made the forest coverage decrease to 32.7% by 1977. Because of vegetation restoration in recent years, the forest coverage has risen to 47.8%, which is still 8.2% lower than that in 1949. In the valleys of the Xiang, Zi, Ruan, and Li rivers in Hunan Province, the forest coverage decreased from 48% in 1957 to 35% in 1984, and the stand volume decreased from 2.8×10^8 to 1.8×10^8 m^3.

1.5.3 The Characteristics of Soil Erosion

The soil erosion in middle and lower Yangtze River has occurred mainly on the cultivated land on slopes, barren-land mountains, and young forestland by water erosion. In addition, in the area of deep weathering layer in Hunan and Jiangxi provinces, landslides and slope disintegration are very common (Photo 1.1).

Red soil, yellow soil, yellow brown soil, and lime soil contain clay, which led to low water permeability and large surface runoff during precipitation. For the purple soil, water permeability is high, but the soil is so thin that it is difficult for reforestation. Without the protection of vegetation, purple soil suffered under heavy precipitation from serious soil losses. The red soil formed by Quaternary red clay is apt to form a loosened top layer, where soil loss occurs easily (Photo 1.2). The red soil formed by shale usually has good structure, which is good for soil erosion control. But when the vegetation was destroyed, soil erosion became serious, causing serious problems for reforestation (Photo 1.3). In the red soil area in southern China, because the red soil and purple soil formed by granite have weak resistance to precipitation, they are easily eroded. The red soil expands in the rainy season and shrinks in dry season. With the alternation of heavy rains and drought, vertical crannies are easily formed, which result in soil erosion.

The hilly area of red soil in the south of Jiangxi Province is an important base of agriculture, forestry, and stockbreeding. In this region, with historically plentiful precipitation and high temperatures, biogeochemical cycling was vigorous, and there was high biodiversity of both plant and animal species. Because of long-term

Photo 1.1 The granite area suffered from serious erosion in Gan prefecture, Anhui Province, China

Photo 1.2 Red soil of the Quaternary suffered from serious erosion in Yiyang prefecture, Jiangxi Province, China

Photo 1.3 Soil of the shale area suffered from serious erosion in Taihe prefecture, Jiangxi Province, China

natural and anthropogenic disturbance, the soil has suffered from serious erosion, because the vegetation is sparse in this region. In some areas, half of the soil in mountainous regions has lost its organic matter, and in locations where the soil was not covered with vegetation, gully erosion and landslides have occurred, leading to the most severely degraded ecosystems (Photo 1.4).

For many years, the governments at all levels and the civilian population have invested much labor, material resources, and financial resources to control soil erosion, such as planting fruit trees, using green manure for vegetation restoration, choosing pioneer species of trees that are resistant to the stress of the barren conditions, adding supplemental nutrients to soil, and planting dryland crops. During the

Photo 1.4 This area without good vegetation cover suffered from a landslide in Taihe prefecture, Jiangxi Province, China

past 50 years, a large area was reforested, leading to natural recovery of communities in the hilly and mountainous regions in southern Jiangxi Province. However, there are still many denuded mountains that need to be managed, which can still be called "difficult places" for reforestation (He, 2003; Zhao, 1995; Xiao, 2002).

References

Abu-Zeid MA (1998) Water and sustainable development: the vision for world water, life and the environment. Water Policy 1:9–19 (in Chinese)

Belt JCB (1975) The 1973 flood and man's constriction of the Mississippi River. Science 189(42040):681–684 (in Chinese)

Du Y (1998) Co-ordinate the relationship between man and nature. J Nat Disasters 7(4):1–6 (in Chinese)

Fang YM, Enroth H, Koponen T, Piippo S (1998) Bryoflora of Jiangxi, China, with an annotated checklist. Hikobia 12:343–364

He F (2003) China central subtropical zone desertification, prevention and control. Soil Water Conserv China 5:15–19 (in Chinese)

Ke JH, Piao SL, Fang JY (2003) NPP and its spatial–temporal patterns in the Yangtze River watershed. Acta Phytoecol Sinica 27(6):764–770 (in Chinese)

Lean J, Warrilow DA (1989) Simulation of the regional climate impact of Amazon deforestation. Nature 342:411–413

Li QK (1983) Chinese red soil. Science Press, Beijing (in Chinese)

Li XW (1996) The statistics analysis of seed plant flora in China. Acta Botanica Yunnanica 18(4):369–384 (in Chinese)

Liu X (1995) Study on seed plant flora in east of China. Acta Botanica Yunnanica 7:93–110 (in Chinese)

Liu SR (1996) Hydrology and ecological function of forest ecosystem in China. China Forestry Publishing House, Beijing (in Chinese)

Luo RY, Gao ZQ (1994) Effects of Chinese fir, loblolly pine and deciduous oak forests on nutrient status of soils in northern subtropics of China. Pedosphere 4(1):1–10

Pielke RA, Dalu G., Snook JS, Lee TJ, Kittle TGF (1991) Nonlinear influence of meso-scale land use on weather and climate. J Climate 4:1053–1069

Qi CJ (1993) Study on plant flora in Badagong Mountain in Hunan province. The corpus of the 66th anniversary of Chinese Botany Academy, Scientific and Technological Publishing House of China, Beijing, 109 (in Chinese)

Rao PC, Enroth H, Piippo S, Koponen T (1997) The bryophytes of Hunan Province, China: An annotated checklist. Hikobia 12:181–203

Redfearn PLJ, Tan BC, He S (1996) A newly updated and annotated checklist of Chinese mosses. J Hattori Bot Lab 69:323–372

Turner MG (1989) Landscape ecology: the effect of pattern on process. Annu Rev Ecol Syst 20:171–197

Xiao HL (2002) Environmental features of Hilly Slope land of red soil in South China and the sustainable utilization. J Mt Res 20(5):32–37 (in Chinese)

Xiong Y, Li QK (1987) Chinese soil. Science Press, Beijing (in Chinese)

Yu YC (1998) Status in soils under different forest stands on the hilly region of southern Jiangsu province. J Zhejiang Forestry Coll 15(1):32–36 (in Chinese)

Yu YC (2000) Effects of continuous plantation of Chinese fir on soil physical properties. J Nanjing Forestry Univ 24(6):36–40 (in Chinese)

Zhang JC, Hu HB (1996) Water and soil loss in China. Chinese Forestry Press, Peking (in Chinese)

Zhang JC, Hu HB, Ruan HH, Fang YM (1999) The present situation and control measures of soil and water loss in the Yangtze River Valley. J Nanjing Forestry Univ 2:18–22 (in Chinese)

Zhao QG (1995) The problem of red soil erosion is no time to delay. Chin Sci Newspaper 6(26):1

Zhou LQ (2005) Relation of flood control with fractural structure, and its activity in the middle and lower reach of the Yangtze river. Yangtze River 36(3):35–38 (in Chinese)

Part II
Development and Application of Soil Loss Models for Soil Loss Prediction in the Shangshe Catchment, Dabie Mountains, China

Chapter 2
Calculation of Water and Sediment Discharge Using an Integral Calculus Method

Abstract In order to improve the accuracy of the calculation of the amount of water discharge, a new integration equation method was developed and applied. In this study, field observation of runoff with a weir was carried out at the outlet of the Shangshe catchments and sub-catchments of four types of land use within the catchment in Yuexi prefecture, Anhui Province. Each weir was equipped with a float-type, water level recorder. In this research, the water level hydrograph of each rainfall–runoff event was divided into periods at 10-min intervals. In each period, a water level versus time linear function was established. Then, the water level versus flux relationship was transferred into a function between flux and time in each period, which could be integrated over time. Thus a new integral equation method was developed to calculate the amount of water discharge in sub-catchments and catchments. The result showed that the new integral equation method can improve the accuracy of annual water and sediment discharge by 8.7–17.1% at five monitoring sites using the field observation data in the year 2000. Compared to the instant flux coefficient method, it improves accuracy greatly and is suitable for the calculation of the amount of water and sediment discharge at weirs under various conditions of water level variation.

2.1 Introduction

During the Yangtze River flood in 1998, the Yangtze's flow peaked at less than 2 million ft^3/s, a rate that it had surpassed 23 times since 1949. Zhang et al. (1999) pointed out that a flow that had caused a small flood in the basin in the past had come to be catastrophic. What would happen if a bigger flow were to occur? Despite the potential hazards to the long-term productivity of arable land due to erosion, which has also led to dam and river bed silting and low soil water retention, very limited information is available about water and soil loss processes from different types of land use. Furthermore, no extensive studies of the problem and its quantification are available. It is easy to calculate the water storage capacity decrease due to roads and construction, and by the shrinking of lakes and reservoirs, but until now we have been unable to quantify the management effects of various types of land use in the mountainous areas on the flood and sediment discharges. Although much research has been carried out concerning water and soil loss rates derived from plot and catchment studies in the Yangtze River basin (Xu, 1999), the results do not provide sufficient information for quantification of the effects of land use on floods in

J.C. Zhang et al., *Theory and Practice of Soil Loss Control in Eastern China*,
DOI 10.1007/978-1-4419-9679-4_2, © Springer Science+Business Media, LLC 2011

this basin. After the 1998 flood of the Yangtze River, recognition of the great importance of paying attention to providing cover for the bare land or very thin forests and shrub-covered land in the mountainous area has become common sense to the environmental experts (Wei and Fu, 1999; Zhang et al., 1999) and the Chinese Forestry Ministry. In 1999, an environmental monitoring project of the Yangtze River protection forest in 11 provinces of the Yangtze River basin was undertaken by the Yangtze River Protection Forest Department of the Chinese Forestry Ministry. That project in the Dabie Mountains of Anhui Province, China, is a main monitoring station.

2.2 Study Area

This study was carried out in the Shangshe catchment (30°12′20″N, 116°26′42″E) of the Dabie Mountains in Yuexi prefecture of Anhui Province, China (Fig. 2.1). The Shangshe catchment comprises an area of 510 ha and has a monsoon climate with a mean annual temperature of 14.6°C and a mean annual precipitation level of 1200 mm. Its non-frost season is 212 days. Concerning soils, according to the Chinese Soil Taxonomy system, the dominant type is mountainous yellow-purple soil derived from weathering granite. This soil has a coarse texture and a high mineral content, which is suitable for cultivation and susceptible to erosion. In this catchment, land types fall mainly into five groups: cultivated land, pine forest, Chinese fir forest, tea garden, and paddy field. The cultivated land types include bare land and roads, the pine forest includes shrubs and bamboo, and the tea garden includes China gooseberry bushes. Within the catchment, the valley and the lowland have been mainly developed into paddy fields. Land with steep slopes is occupied by pine forest and Chinese fir forest. Most cultivated land can be found at the foot of the mountains or hills near the valley. Land use information, obtained from a 1:10,000 scale land use map (made by the Forestry Department of Yuexi County, Anhui Province), indicates the following distribution (Fig. 2.2): pine forest comprises an area of 293.8 ha, which takes up 55% of the Shangshe catchment; paddy field is the second largest, comprising an area of 78.4 ha taking up 14.8% of the total

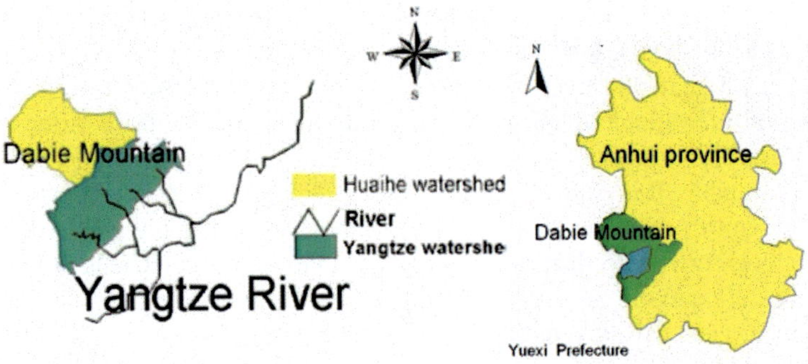

Fig. 2.1 Dabie Mountains of Yangtze River basin in Anhui Province, China

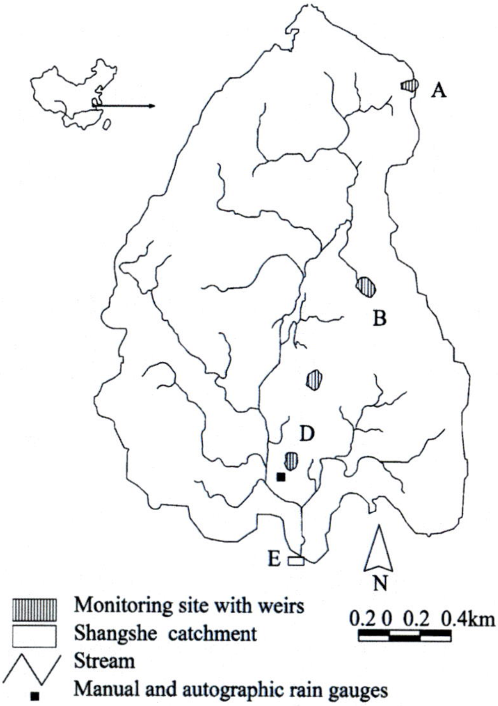

Fig. 2.2 Location of the research site and the gauges in the Shangshe catchments. A is the sub-catchment of the tea garden (0.59 ha); B is the sub-catchment of Chinese fir forest (0.89 ha); C is the sub-catchment of pine forest (0.86 ha); D is the sub-catchment of cultivated land (0.74 ha); E is all of the Shangshe catchment (510.0 ha)

area. The areas of cultivated land, Chinese fir forest, and tea garden are 44.1, 45.9, and 65.8 ha, taking up 8.4, 8.7, and 12.5%, respectively.

The main tree species are masson pine (*Pinus massoniana*), Chinese fir (*Cunninghamia lanceolata*), bamboo (*Phyllostachys edulis*), oriental white oak (*Quercus alba*), chestnut (*Castanea mollissima*), Chinese blue oak (*Cyclobalanopsis glauca*), dye tree (*Strobilacea platycarya*), and Chinese hackberry (*Celtis tetrandra*). Among these, masson pine and Chinese fir make up the greatest fraction. Sweetleaf sapphireberry (*Symplocos paniculata*), Sims Azalea (*Rhododendron simsii*), oil Camellia (*Camellia oleifera*), rubus (*Rubus palmatus*), and Indian strawberry (*Duchesnea indica*) are the main kinds of shrub.

2.3 Materials and Methods

2.3.1 Water Runoff Observation

A rectangular weir made of concrete and a 90° sharp-crested V-notch weir equipped with a float-type water level recorder were built at the outlet of the Shangshe catchment (Fig. 2.2). In the Shangshe catchment, four small catchments of different land types were selected as experiment sites (Fig. 2.3, Photo 2.1).

Fig. 2.3 Land use map of the Shangshe catchment

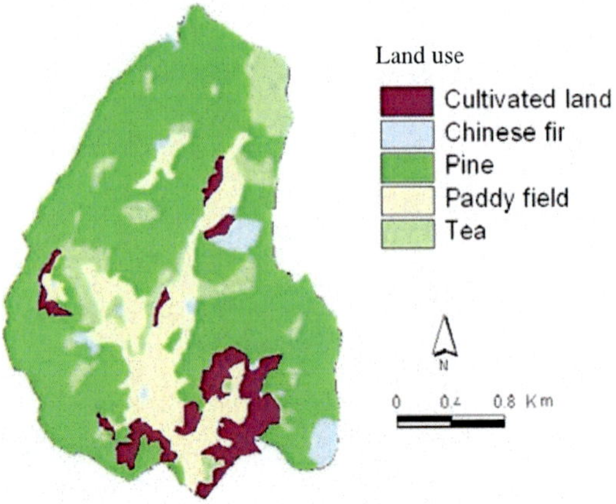

Four 45° sharp-crested, V-notch weirs were set up in the catchments of a pine forest (0.86 ha), a Chinese fir forest (0.89 ha), cultivated land (0.74 ha), and a tea garden (0.59 ha) (Photo 2.2). The main human activity in cultivated land is the clear-cutting of wheat in June, the cultivation of land in early July for corn planting, and

Photo 2.1 Monitoring gauges in the Shangshe catchment: (**a**) the river outlet of the Shangshe catchment; (**b**) a rectangular weir made of concrete and a 90° V-notch weir at the river outlet are built, equipped with a float-type water level recorder; (**c**) a manual rain gauge and an automatic rain gauge 400 m away from the river outlet in the Shangshe catchment

Photo 2.2 Sub-catchments of various types of land use with monitoring gauges

fertilization in mid-August. Clear-cutting of grass in March and August was carried out in the tea garden, and the remains of the grass were left on the terraces as strip covers. Selected characteristics of the five monitoring sites are listed in Table 2.1. Every morning at 8 o'clock, the recorder paper of each water gauge was changed

Table 2.1 Characteristics of the five monitoring sites

Monitoring site	Area (ha)	Soil depth (cm)	Cover	$\overline{H}$ (m)	$\overline{D}_{1.3}$ (cm)	Density of individual (N ha^{-1})
Chinese fir forest	0.89	35	Fir tree and shrub	8.3	12.1	1725
Pine forest	0.86	45	Pine tree and shrub	6.2	10.3	1250
Cultivated land	0.74	16	Wheat and oil seed rape from Jan to June and corn from June to Oct	Wheat 0–1 m and Corn 0–2.5 m		
Tea garden	0.59	45	Tea and grass	0.5		
River outlet	510					

$\overline{H}$ is the average height of the trees and crops in the monitoring sites (m); and $\overline{D}_{1.3}$ is the average diameter of trees at the height of 1.3 m (cm)

manually, and the water level of each weir was measured to be used to correct the error of the water level in the recorder paper. Field observation data of water runoff at the five monitoring sites in the years 1999 and 2000 were used in this study

2.3.2 Precipitation Observation

A manual rain gauge and an autographic rain gauge were set up to survey precipitation and rainfall intensities 100 m away from the weir at the monitored site of cultivated land and 400 m from the weir at the monitored site of the pine forest. Every morning at 8 o'clock the recorder paper of the autographic rain gauge was changed. For every rainfall event the precipitation was read from the recorder paper of automatic rain gauge and compared with the data from the manual rain gauge. Precipitation data for 1999 and 2008 obtained from the autographic rain gauge were used in this study. Precipitation data for 2001 and 2008 obtained from the manual rain gauge were also used.

2.3.3 Suspended Sediment Observation

A simple laboratory in the Shangshe catchment was equipped with an oven and electric scale. For each rainfall event in year 2000 that led to runoff, 500 ml of water was collected manually at intervals of 10 min from each weir and labeled with time, place, and water level. Then organic matter, such as grass and leaves, was picked out. From each of these samples, 30 ml was taken out. Sands were filtered through filter paper of 0.2 μm and put into the oven at 110°C for 24 h, to evaporate the water and measure the SC (sediment concentration, kg m^{-3}). According to Mihara (1996), who studied the effects of agricultural land consolidation on erosion processes in semi-mountainous paddy fields of Japan, when the paddy field surface was flooded, leveling of adjacent surfaces and construction of levee slopes were essential, and erosion in paddy fields occurred mostly on levee slopes. Lu and Higgitt (2001) studied temporal changes in sediment yields in the Yiwanshui catchment in western China by using caesium-137 (^{137}Cs) dating techniques and found the reservoir to contain as much as 84% of the eroded soil, with the remaining 16% being attributed to valley-floor paddy fields. In our research, samples taken from the flushed paddy field showed similar SSC (suspended sediment concentration) between the flush in and flush out of the paddy field. Accordingly, soil erosion and sedimentation in the paddy field were not considered in this study.

Photo 2.2a is the sub-catchment of cultivated land, which encompasses an area of 7400 m^2. The crops in cultivated land are wheat from January to June and corn from the end of June until October. The main cultivation measures are carried out at the middle term of June for corn planting, after clearing of the wheat crop, and December for wheat planting. Fertility measures are carried out mainly in August with weeding and soil tilling. Photo 2.2b is the sub-catchment of pine forest with an average age of 20 years, which encompasses an area of 8600 m^2. Photo 2.2c is the sub-catchment of tea garden, in which tea was planted after pine forest was cut in 1997. The catchment of the tea garden encompasses an area of 5900 m^2. Grass-clearing measures are carried out in March and August and the remains of the cut grass are put on the terraces as strip covers. Photo 2.2d is the sub-catchment of 16-year-old Chinese fir forest, which encompasses an area of 8900 m^2.

Field observation data of precipitation, sediment discharge, and runoff at different scales from micro-plot, USLE (universe soil loss equation) plot, sub-catchment to catchment scale were used in this study in 10 years from 1999 to 2008. Field observation data of sediment discharge and runoff at micro-plot and USLE-plot scales in each type of land use will be introduced in Chapters 4, 5, and 6, respectively.

2.4 Method to Calculate the Amount of Water and Sediment Discharge

2.4.1 Method to Calculate the Amount of Water Discharge

For many years in China, the standard method to calculate the amount of runoff has been to use an averaging method in hydrological experiments (Fu, 1998) as

$$Q = \sum_{i=1}^{n} ((Q_{s,i-1} + Q_{s,i})/2)t_i \tag{2.1}$$

where

Q is the amount of water discharge in a period (m^3);
$Q_{s,i-1}$ and $Q_{s,i}$ are the fluxes of water discharge based on a $H - Q_s$ relationship at the beginning and the end of a period, respectively (m^3/s);
H is the water level (m);
and t_i is the time in the period (s).

However, an averaging method is most suitable for cases of small changes of water level at a weir, or of flow on an even keel. In the Dabie Mountains, it is suitable for the study of cases of runoff produced by small rainfall events. In the case of a storm, however, water levels of surface runoff of a triangular weir in forest and sloping land respond to rainfall rapidly, in only a few minutes. Because of the great change of water level in a short period, the averaging method will necessarily result in large errors in calculating the amount of water discharge. Figure 2.4 presents an example of the water level (H, m) – Q_s (m^3/s) relationship of a 45° sharp-crested V-notch. It also shows a comparison between the methods of calculating the amount of water discharge in a certain period using Eq. (2.1), along with an integral calculus method. The triangular ABC in Fig. 2.4 represents the water discharge calculated with Eq. (2.1). The shade of ABC in Fig. 2.4 represents the water discharge calculated with an integral method.

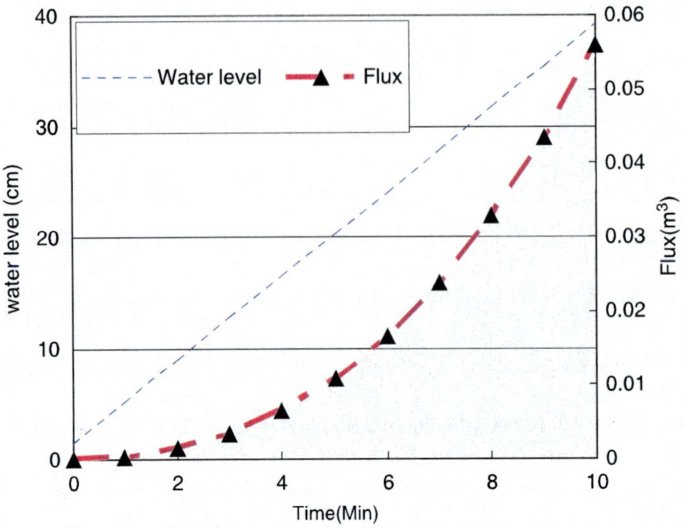

Fig. 2.4 Comparison of two methods to calculate the amount of water discharge

The integral calculus method, which can obviate the shortcoming of Eq. (2.1), is suitable for all cases of the calculation of amount of water discharge with high accuracy (Fig. 2.4).

This method is as follows:

First, the water level line is divided into periods at 10-min intervals. Then the beginning water level H_A and the ending water level H_B in each period are recorded. The equation of the flux according to the water level of a triangular sharp-crested weir is as follows (Dong, 1995):

$$Q_s = m_0 b \sqrt{g} H^{2.5} \tag{2.2}$$

where

Q_s is the flux of water discharge (m^3/s);
m_0 and b are parameters of the weir;
the variable g represents the acceleration of gravity; and
H is the water level (m).

When the sharp-crested weir has a 45° degree roof angle, the equation to calculate the flux is

$$Q_s = 0.58247 H^{2.5} \tag{2.3}$$

In a 90° case, the equation to calculate the flux is

$$Q_s = 1.4 H^{2.5} \tag{2.4}$$

where

Q_s is the flux of water discharge (m^3/s) ; and
H is the water level of a triangular weir with 45° roof angle (m).

Using integral calculus from Eq. (2.3), Eq. (2.5) is obtained:

$$\int Q_s \, dt = \int 0.58247H^{2.5} \, dt \tag{2.5}$$

Because in any one interval the water level line at the weir is straight, or nearly straight, Eq. (2.6) pertains:

$$H_t = H_A + ((H_B - H_A)/t_i)t \tag{2.6}$$

where

H_t is the water level at any time of the AB period (m);
H_A is the water level at the beginning of the AB period (m);
H_B is the water level at the end of the AB period (m);
t is the time during the AB period (s); and t_i is the total time of AB period (s).

In Eq. (2.6), $(H_B - H_A)/t_i$ is the slope of a line. Let this be represented by K; then Eqs. (2.7) and (2.8) are obtained:

$$\int Q_s \, dt = \int 0.58247H^{2.5} \, dt = \int 0.58247(H_A + Kt)^{2.5} \, dt$$

$$\int Q_s dt = 0.58247 \times (2/7)(1/K)(H_A + Kt)^{3.5} \tag{2.7}$$

$$\int_0^t Q_s \, dt = 0.58247 \times ((2/7)(1/K)(H_t{}^{3.5} - H_A{}^{3.5})) \tag{2.8}$$

Similarly, Eq. (2.9) is used to calculate the flux for a 90° sharp-crested weir:

$$\int_0^t Q_s dt = 1.4 \times ((2/7)(1/K)(H_t{}^{3.5} - H_A{}^{3.5})) \tag{2.9}$$

For a water level over the crested weir at the river outlet, which is made of a 90° sharp-crested weir and a wide concrete weir, the total flux consists of two parts: the flux from the 90° sharp-crested weir and that from wide concrete weir. Eq. (2.10) (Dong, 1995) is used to calculate the flux:

$$\int_0^t (Q_s + 0.26)dt = \int_0^t a(bH + H^{1.5})dt + \int_0^t 0.26 \, dt \tag{2.10}$$

where

Q_s is the flux from the concrete weir (m^3/s);
the value 0.26 represents the flux of water from the 90° sharp-crested weir, when the water level begins to spill over the weir (m^3/s); and
a and b are parameters of the concrete weir, which are 25.734 and 0.933, respectively.

Figure 2.4 depicts the difference between the amount of runoff calculated using the averaging method and the integral calculus method. The amount of water discharge calculated using the averaging method is higher than that by the integral calculus method. In addition, the error increases with the variance of water level. For example, Table 2.2 shows the example of a storm on 15 Aug. 2000. When the water level varied from 9.3 to 39.2 cm in a 10-min interval, the amount of water discharge calculated by the averaging method was 9.96 m^3, which was 68.8% more than that calculated by the integral method. The total error on this day was 15.67 m^3. Table 2.3 shows that the integral calculus method can improve the accuracy of annual amount of water discharge by 8.7–17.1% at the five monitoring sites using the data in the year 2000.

Table 2.2 Comparison between values of water discharge calculated by two methods in the sub-catchment of cultivated land on 15 Aug. 2000

Date	Time (min)	Rain (mm)	Water level (cm)	Q_s (m^3/s)	Q_{av}(m^3)	Q_{int} (m^3)	Error	%
Aug. 15th	0		1.4	0.00000	0.00	0.00	0.00	0.00
	10	9.3	39.2	0.05604	16.82	9.96	6.86	68.84
	20	9.5	51.3	0.10979	49.75	48.63	1.12	2.29
	30		18	0.00801	35.34	28.26	7.08	25.04
	40		5	0.00033	2.50	1.88	0.62	32.93
Total					104.4	88.74	15.67	12.6

Q_{av} is the amount of water discharge of a period calculated using the averaging method (m^3) and Q_{int} is the amount of water discharge of a period calculated using the integration formula (m^3)

Table 2.3 Comparison between annual fluxes calculated by the two methods of various land use in the year 2000

Land use	Cultivated land	Chinese fir forest	Pine forest	Tea garden	River outlet	
Q_{av} (m^3)	2445.4	911.5	932.7	956.7	902, 123.5	$\Delta t = 10$ min
Q_{int} (m^3)	2211.7	797.1	796.8	872.2	830, 067.3	
Error (m^3)	233.8	114.4	136.0	84.5	72, 056.2	
%	10.6	14.4	17.1	9.7	8.7	

Q_{av} is the amount of water discharge of a period calculated using the average method (m^3) and Q_{int} is the amount of water discharge of a period calculated using the integration formula (m^3)

2.4.2 *Method to Calculate Sediment Load*

Equation (2.11) is used to calculate the sediment discharge at the sub-catchment and catchment scale:

$$\mathrm{SD} = \sum_{i=1}^{n} \mathrm{SD}_i = \sum_{i=1}^{n} \overline{C_i} Q_{fi} \tag{2.11}$$

where

SD is the sediment discharge in the whole period (kg);
SD_i is the sediment discharge in the ith period (kg);
$\overline{C_i}$ is the average suspended sediment concentration in the ith period (kg m^{-3}); and
Q_{fi} is the amount of water discharge calculated by integral equation in the ith period (m^3).

The sediment discharge was transformed into specific sediment discharge using the following equation:

$$\mathrm{SSD} = \mathrm{SD}/S_{\mathrm{A}} \tag{2.12}$$

where

SSD represents specific sediment discharge (t km^{-2});
SD is the sediment discharge from each sub-catchment or catchment (ton, t); and
S_{A} is the area of each sub-catchment or catchment (km^2).

Because the sediment discharge was calculated with the same SSC data at each monitoring sites in the Shangshe catchment, application of the integral calculus method can also improve the accuracy of the annual amount of sediment discharge by 8.7–17.1%.

2.5 Comparison of Soil Loss Among Various Types of Land Use

Using the USLE-plot field observation data in the Shangshe catchment, annual soil loss from different land types (pine forest, Chinese fir forest, cultivated land, and tea garden) was calculated (Fig. 2.5). Figure 2.5 shows farmland that was planted with wheat from November to June in the second year and corn from June to November, of which during the growing season, the coverage was more than 80%. During the harvest–planting–seedling period (after mid-June to early August and early November to late December), the plant cover is close to 0. In the year 2000 in the planting–seedling period of corn (June through August), because of highly

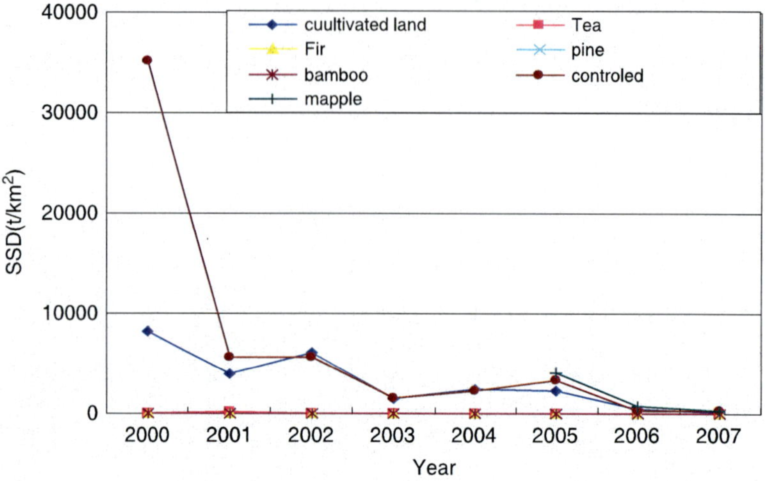

Fig. 2.5 Comparison of soil loss among seven types of land use in the years 2000–2007

intense precipitation, soil erosion was very serious in the area, with runoff and erosion occurring when daily rainfall reached close to 10 mm. Soil loss in the year 2000 amounted to 35,000 t km^{-2}. In the years of 2001 and 2002, because in the planting–seedling period, rainfall was reduced by 30–50% compared to that of the year 2000, soil erosion was correspondingly reduced by the amounts of 30–50%. In year 2003 a project of returning farmland on slopes to forests started in the sub-catchment of cultivated land. Poplar trees and ginseng were planted. Then a large decline in soil erosion was shown. The status of the sub-catchment of cultivated land before and after "the returning farmland to forest project" is shown in Photos 2.3, 2.4, and 2.5. Figure 2.5 shows that in the year 2005, despite heavy rainfall, the annual amount of soil erosion was still less than 4000 t km^{-2}. In the years 2006 and 2007, the amount of soil erosion dropped to 504 and 309 t km^{-2}, respectively. The effects of soil erosion control increased with the coverage of mulberry trees (*Morus alba*) and poplar trees (*Populus euramericana*) (Photo 2.5), which is very significant. In year 2005, in the runoff of a plot of mixed young Formosan sweet gum (*Liquidambar formosana*) and pine afforestation, due to higher rainfall erosivity, soil erosion reached 4000 t km^{-2}. In years of 2006 and 2007, due to the percentage of vegetation coverage increasing to 45 and 70%, on-site runoff and soil loss decreased significantly (Fig. 2.6) so that soil loss was reduced to 774 and 305 t km^{-2}, respectively.

Soil erosion in the sloping land of a mixed young maple and pine plot and the control plot was much higher than that of tea, Chinese fir, pine forest, and bamboo forest. In order to better explain soil erosion from tea garden, Chinese fir forest, pine forest, and bamboo forest, in the years of 2000–2007, a detailed soil erosion comparison of four types of land use was made (Fig. 2.6). Figure 2.6 shows that soil erosion in the year 2001 was 55 t km^{-2} in the tea garden, much higher than other years (which was less than 10 t km^{-2}), mainly due to the grass cutting and

Photo 2.3 Sub-catchments of cultivated land monitoring site before reforestation (1999.8)

Photo 2.4 Sub-catchments of cultivated land monitoring site before reforestation (2003.3)

tending during the rainy season. For the Chinese fir forest, pine forest, and bamboo forest, the annual soil losses were less than 10 t km^{-2}, except for the pine forest in 2005, the annual amount of which was 10.9 t km^{-2}. Soil loss in the forest was very low because the multi-layer vegetation of forest intercepted rainfall, weakened the kinetic energy of the raindrops, and decreased the direct impact of the raindrops on

Photo 2.5 Sub-catchment of cultivated land monitoring site after reforestation (2008.7)

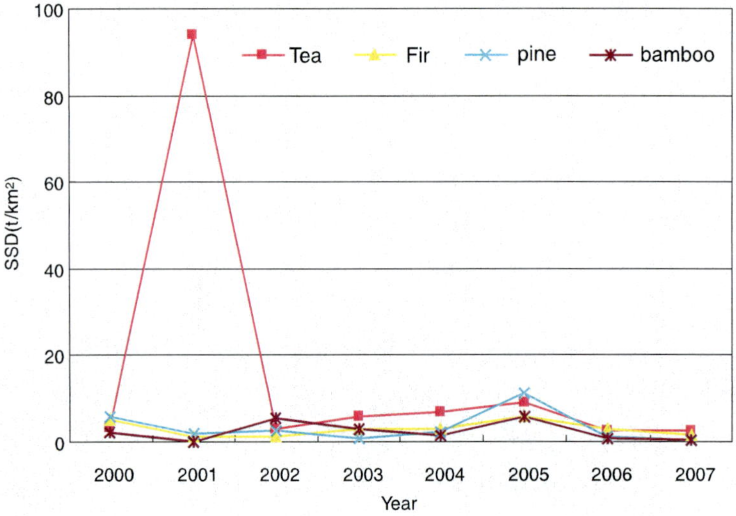

Fig. 2.6 Comparison of soil loss among seven types of land use in the years from 2000 to 2007

soil. At the same time, due to better soil structure for infiltration, and particularly due to the litter, a considerable part of the surface runoff was absorbed into the ground floor. The small amount of sediment loss of forests was also due to the well-developed root system for soil erosion control.

Photo 2.6 A USLE plot of maple cross pine tree (2008.7)

Photo 2.7 Pine seeding entered the control plot and grew well (2008.7)

Photo 2.8 Students of soil and water conservation of Nanjing Forestry University were making investigation of pine biomass in the control plot (2008.8)

In years 1999 and 2000, the bush in the control plot became bare land. Serious soil erosion occurred when daily precipitation reached 7–8 mm. Soil loss in the year of 2000 amounted to 35,000 t km^{-2}. After 2001 natural regeneration of vegetation gradually occupied the controlled plot and coverage increased to 70 and 80% (Photos 2.5, 2.6, 2.7, and 2.8) by the years of 2006 and 2007, respectively, the amount of soil erosion decreased significantly to 305 and 303 t km^{-2}.

References

Dong ZN (1995) Hydrology. Higher Education Press, Beijing

Fu J (1998) Hand book for carrying out hydrological and environmental field observations in Anhui Province. Department of Forestry, Hefei, China

Lu XX, Higgitt DL (2001) Sediment delivery to the three gorges. Geomorphology 41:157–169

Mihara M (1996) Effects of agricultural land consolidation on erosion processes in semi-mountainous paddy fields of Japan. J Agric Eng Res 164:237–247

Wei HL, Fu CB (1999) Study of the sensitivity of a regional model in response to land covers change over Northern China. Hudro Process 12:2249–2265

Xu Y (1999) Flood hazards in the Yangtze River Basin and countermeasure. Academic Press, Beijing, p 273 (in Chinese)

Zhang JC, Ruan HH, Hu HB et al (1999) The present situation and control measures of soil and water loss in Changjiang River Valley. J Nanjing Forestry Univ 23(2):18–22

Chapter 3
Development of the GOIUG Model with a Focus on the Influence of Land Use in the Shangshe Catchment

Abstract Sediment management strategies are crucial to developing countries because of the limited resources. Establishment of these strategies is hampered to some extent by the lack of reliable information on catchment sources. This chapter reports the rainfall-sediment response of various types of lands in the Shangshe catchment (528.8 ha) of the Dabie Mountains of Anhui Province, China. Field observation in the year 2000 showed 32 events occurred, among which 18 events produced runoff. Comparison among SISSD (specific instant suspended sediment discharge) hydrographs from the five monitored sites showed that the cultivated land produced the highest level of SISSD, which was 100 times higher than that of the tea garden, Chinese fir forest, or the pine forest, and 10 times higher than that of the river outlet. Integrating the field observation data of SISSD in representative kinds of land use within the catchments with geographic information system (GIS) and a hydro model, the GOIUH (GIS-based observed instantaneous unit hydrograph) model, was constructed to simulate the SISSD hydrograph of the river outlet. The resulting model showed that the calculated SISSD hydrographs compared well with the observed ones at the outlet of the Shangshe catchment, as shown by a coefficient of correlation R^2 of 0.8. The semi-distributed sediment discharge model, with a focus on the influence of land use in the Shangshe catchment, can help to quantitatively understand the source of sediment discharge.

3.1 Introduction

Sediment transport in mountain streams, which normally occurs during single floods, displays a great variability from event to event, as well as intra-event (Mario and Lorenzo, 2000). It has been known for a long time that the relationship between water discharge and sediment concentration in rivers shows important variation at different timescales. The relationship between sediment concentration and stream discharge varies among seasons due to the development of the crops and the influence of discharge (Gregory and Walling, 1973; Edmonds, 1992). On a global scale, the importance of topography and the significance of relatively small mountainous catchments as major contributors to global continental sediment discharge has been noted (Milliman and Syvitski, 1992). The difficulty of obtaining sufficiently detailed, spatially distributed data on catchment characteristics has hampered attempts to disentangle the various controls on sediment yields within large catchments (Yang et al., 2002). However, the availability of global environmental

J.C. Zhang et al., *Theory and Practice of Soil Loss Control in Eastern China*,
DOI 10.1007/978-1-4419-9679-4_3,

data sets comprising description of hydroclimatic, biological, and geomorphological characteristics of the earth offers a means of extracting catchment variables for integration with sediment yield data. This approach has been used to examine global variations in sediment yield where individual catchments are represented by a single sediment yield value (Summerfield and Hulton, 1994). In the present study this method is extended to the investigation of sediment yields within the Shangshe catchment of the Dabie Mountains of the lower Yangtze basin, China. Analysis of suspended sediment discharge related to types and locations of active sediment sources has been previously reported (e.g. Xiang and Zhou, 1986). However, a better understanding of concentration–discharge relationships will be possible only if an explicit linkage can be established between the sediment export at the outlet and the spatial distribution and rates of processes within the catchments (Walling, 1996). Among the various studies reported to establish this linkage, the GIS-based hydrology and sediment models are powerful tools in support of watershed analysis and assessment of management scenarios at the watershed scale.

Hydrologic simulation began in the 1950s with the advent of the digital computer. Among the various applications of hydrologic simulation models are streamflow forecasting, design and planning, and extension of stream flow records (Pascal and Laura, 1998). The first models, of which the Stanford Watershed Model of Crawford and Linsley (1966) may be one of the best known examples, were spatially lumped, meaning that the models represented the effective response of entire catchment, without explicitly attempting to characterize spatial variability of the response. Spatially lumped stream-flow simulation models are still widely used, but a critical shortcoming of the spatially lumped stream-flow simulation models is their inability to represent the spatial variability of hydrologic processes and catchment parameters (Moore and Grayson, 1991). In recent years, a second generation of hydrologic simulation models have been developed, which are able to spatially represent some, if not most, important land surface characteristics.

In addition, sediment discharge modeling, which is closely related to hydrologic simulation, has also been developed. Many distributed sediment discharge models have used the method of combining hydrological models with soil erosion models, such as LISEM (De Roo et al., 1996), WaSiM-ETH-AGNPS (Rode et al., 2000), and DXAJ (Su et al., 2003). They have been developed with a coupled structure, i.e., algorithms describing hydrological process form the fundamental components in the model, with which algorithms associated with soil erosion and sediment discharge models have been coupled (Rudra et al., 1995; Su et al., 2003).

In theory, detailed information on slope profile and land use, incorporated into a GIS-based model, should lead to better results than those obtained when a lumped approach is used for hydro and sediment discharge modeling. At present, however, physically detailed sediment discharge models do not necessarily produce better results than do much simpler and partially lumped sediment discharge models, although the former represent topography in more detail. This was demonstrated by De Roo (1993) and Jette and De Roo (1998) when simulation results from several runoff and sediment discharge models were compared with measured water and sediment discharge. Even now, quantitative understanding and prediction of the process of runoff and suspended sediment generation and its transmission to the outlet represents one of the most basic challenges in hydrology (Jain and Singh, 2000). This chapter reports the results of a study which was set up to examine the linkage between sediment export and intra-basin processes in the Shangshe catchment of the Dabie Mountains. Sediment discharge was monitored at the river outlet of the Shangshe catchment and the four small water-collecting areas of various land types in it. The main objectives of this study were the following: (1) to detect variations in the *SISSD* (specific instantaneous suspended sediment discharge) graph according to land use and (2) to simulate the *SISSD* graph at the river outlet of the Shangshe catchment using a spatially semi-distributed sediment discharge model (developed using the unit-graph method by integrating the observed *SISSD* graphs of representative monitored sites of various land types with GIS data and a hydro model).

3.2 The Study Area

This study was carried out in the Shangshe catchment (34°32′20″*N*, 116°50′12″*E*) of the Dabie Mountains in Yuexi prefecture of Anhui Province, China (Fig. 2.1). Details of the Shangshe catchment are reported in Chapter 2.

3.3 Materials and Methods

3.3.1 Water Runoff Observation and Precipitation

Details of the water runoff observation at five monitoring sites with weirs and precipitation observations are reported in Chapter 2.

3.3.2 Suspended Sediment Observation

Details of suspended sediment observation during the year 2000 at five monitoring sites with weirs are reported in Chapter 2. For some events, samples were taken at intervals of 20 min or longer. No samples were taken over the period in which the water level varied slowly, and the last sample taken near that period was applied as an estimate for that period.

Observational data from the year 2000 were selected for this study. In the year 2000, there were 78 rain events, among which 18 events produced runoff. During the 18 rainfall–runoff events, *SSC* (suspended sediment concentration) was measured at intervals of 10 min at the five monitored sites. The *SSC* (kg m^{-3}) was transformed into *SISSD* (kg km^{-2} s^{-1}) using the following equation:

$$\mathrm{SISSD} = Q_{\mathrm{s}}/S_{\mathrm{A}} \times \mathrm{SSC} \tag{3.1}$$

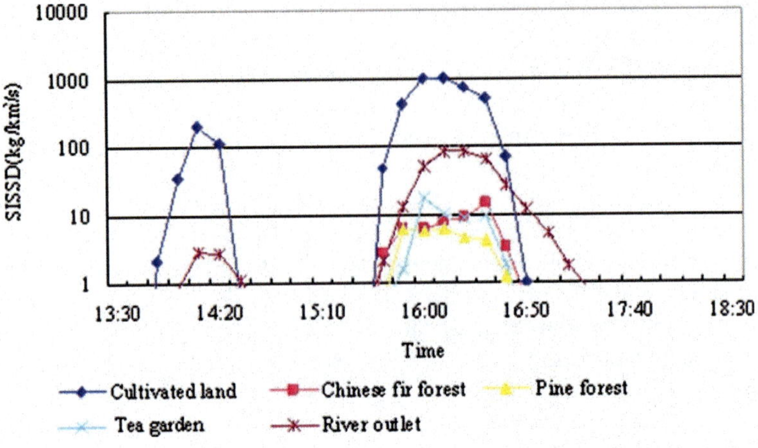

Fig. 3.1 *SISSD* graphs of the five monitored sites on 18 August 2000

where

SISSD is the specific instantaneous suspended sediment discharge (kg km^{-2} s^{-1});
Q_s is the observed flux of water discharge calculated from the H–Q_s relationships (m^3 s^{-1});
S_A is the area of the catchment with weir (km^2); and
SSC is the suspended sediment concentration sampled at given values of Q_s (kg m^{-3}).

SISSD graphs were made for every rainfall-sediment discharge event at every monitored site in the year 2000. In a monsoon climate, precipitation mainly occurs during the summer in storms; runoff and sediment discharge are mainly produced by storms from May to September. The event on 18 Aug 2000 showed the typical characteristics of a storm, in terms of runoff and suspended sediment discharge, at the five monitoring sites, especially in the cultivated land 2 weeks after clear-cutting of wheat and a few days after tillage for corn seeding. Here an example of the 18 events is shown in Fig. 3.1.

Figure 3.1 shows the *SISSD* graph at the five monitoring sites on 18 August 2000. The *SISSD* graphs at the five monitored sites all showed correspondence to the rainfall hyetograph. The *SISSD* graph for cultivated land was the highest among the five sites and its peak (1022 kg km^{-2} s^{-1}) was about 18 times that of the river outlet site (83.3 kg km^{-2}s^{-1}). The peaks of *SISSD* in the pine forest, Chinese fir forest, and tea garden were low (less than 20 kg km^{-2} s^{-1}), about one-fiftieth of that of the cultivated land. The differences in the *SISSD* peaks among pine forest, Chinese fir forest, and tea garden were small. These values showed similarities with the results of Jean (2003), who reported the effects of human activities on SSC during the late 20th century at 57 sites along major rivers in central Japan using governmental data and GIS.

3.3.3 GOIUG Model

The unit graph method is a simple method of rainfall–runoff modeling. A significant advance in the unit graph method was the development of the geomorphological instantaneous unit hydrograph (GIUH) (Lee, 1998). The GIUH approach was originated by Rodriguez-Iturbe and Valdes (1979), who reasonably interpreted the runoff graph within the framework of "concentration time" distribution (distribution of travel times from points in a catchment to the outlet) for the geo-morphological structure of a basin. In this chapter, by interpreting the observed *SISSD* graphs of various land types in the framework of a concentration time distribution, a new formula has been developed for the calculation of the *SISSD* graphs at the river outlet, which may be called GOIUG method (GIS-based observed instantaneous unit graph), developed from the GIUH method. However, unlike the GIUH method, in which the runoff hydrograph is calculated by a mathematical method, the GOIUG method aims to simulate the *SISSD* graphs at the river outlet using the observed *SISSD* graphs of representative kinds of land use within the catchments introduced above, for each rainfall-sediment discharge event.

The key characteristic of this method is that it uses the observed *SISSD* graphs as the basic unit graphs in each rainfall event for corresponding kinds of land use. Consequently there are only a few parameters in the model. The observed *SISSD* graphs, which show the real response of rainfall-sediment discharge of various kinds of land use, are used as the basis and are believed to be able to accurately simulate the *SISSD* graphs at the river outlet.

In this method, field observations of the *SISSD* graphs were carried out in each representative catchment of different kinds of land use. The geomorphology of the monitored sites was taken as representative of every corresponding kind of land use within the Shangshe catchment. For each observed *SISSD* graph, it was presumed that the same land use has the same rainfall-sediment discharge response. These responses were then applied to other places to represent the rainfall-sediment discharge response to the same rainfall event in corresponding types of land use, despite the fact that the coverage, slope, and soil characteristics might be different from those of the corresponding kinds of the monitored sites.

For each rainfall-sediment discharge event in each kind of land use, the observed *SISSD* graph was taken as a unit. Combining the observed *SISSD* graphs of each monitored site in each rainfall-sediment discharge event with the concentration time and area of various land types within each interval of concentration time allowed the *SISSD* graph at the river outlet to be simulated. The calculation method of the GOIUG is shown in Fig. 3.2 as an example.

Figure 3.2b shows the two catchments (C_1 and C_2) with the same outlet B. Figure 3.2a shows that the observed *SISSD* graph was used as a basic unit to calculate the *SISSD* graphs of the C_1 and C_2 catchments, as well as the graph of *ISSD* (instantaneous suspended sediment discharge, kg s^{-1}) of the B catchment, considering the concentration time and the area of various land types within each interval of concentration time.

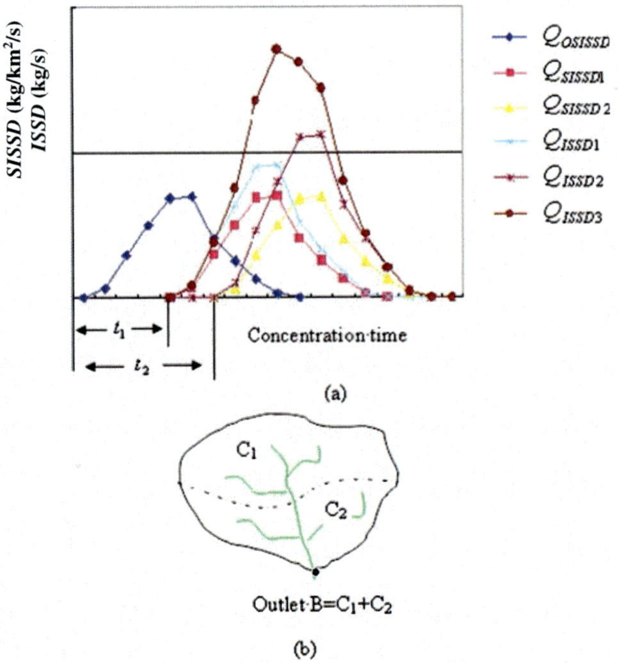

Fig. 3.2 An example of the GOIUG method to calculate the graph of sediment discharge at the river outlet. Q_{OSISSD} is the observed *SISSD* graph of one experimental site with a weir (kg km^{-2} s^{-1}); Q_{SISSD1} is the *SISSD* graph of the C_1 catchment, calculated by moving the graph of Q_{OSISSD} along the *X*-axis through t_1 units of time (kg km^{-2} s^{-1}); Q_{SISSD2} is the *SISSD* graph of the C_2 catchment, calculated by moving the graph of Q_{OSISSD} along the *X*-axis through t_2 units of time (kg km^{-2} s^{-1}); Q_{ISSD1} is the calculated *ISSD* graph from the C_1 catchment at the river outlet (kg s^{-1}); $Q_{ISSD1} = Q_{SISSD1}S_1$, where the S_1 is the area of the C_1 catchment (km^2). Similarly, Q_{ISSD2} is the calculated *ISSD* graph from the C_2 catchment at the river outlet (kg s^{-1}), $Q_{ISSD2} = Q_{SISSD2}S_2$, where S_2 is the area of the C_2 catchment (km^2); Q_{ISSD3} is the total of the calculated ISSD graph from the B ($C_1 + C_2$) catchment (kg s^{-1}), $Q_{ISSD3} = Q_{ISSD1} + Q_{ISSD2}$

The water flow length was calculated with the hydrological model module under ArcView, a GIS software package from ESRI (Redland, USA). At first, scanned raster data from elevation of the Shangshe catchment were loaded into the 3D module of ArcView to derive a digital elevation model (DEM) of the catchment (Fig. 3.3). Then the stream networks and catchments were delineated from the DEM using the hydrological model module of ArcView (Fig. 3.4). The catchments' properties were derived within the GIS environment using the polygons of the delineated catchments. The water flow length of the Shangshe catchment is one of the most important hydrological properties and is shown in Fig. 3.5.

The upslope elevation and the downslope elevation of the main streams in each catchment were calculated within the GIS environment to calculate the average slopes of the streams. In this study, concentration time is the travel time of

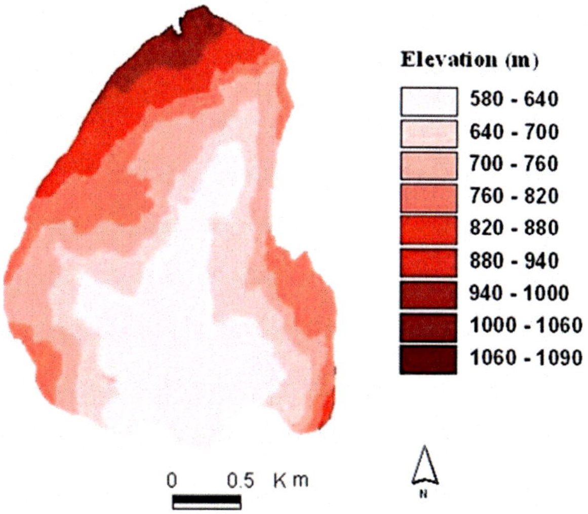

Fig. 3.3 DEM of the Shangshe catchment

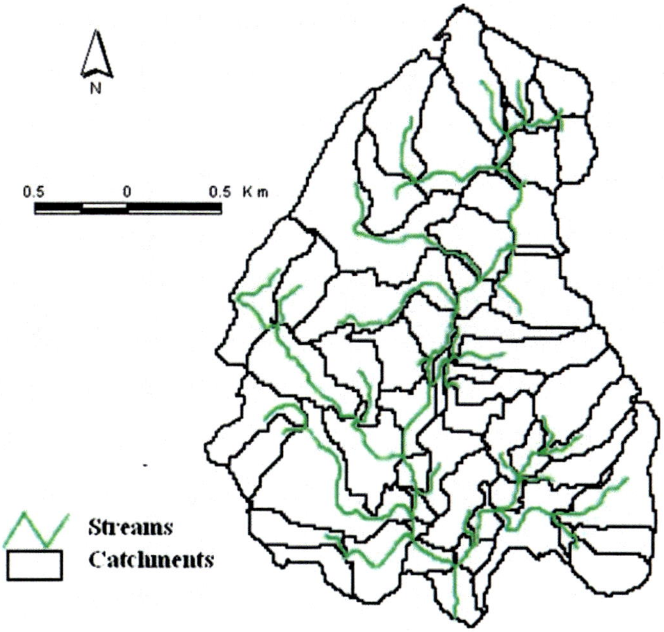

Fig. 3.4 Small catchments and streams in the Shangshe catchment

the *SISSD* graph of any point within the Shangshe catchment from its source to the river outlet. A time–area diagram is prepared considering the traveling time in any small catchment proportional to the distance and inversely proportional to the square root of the average slope of the main stream within it. An initial

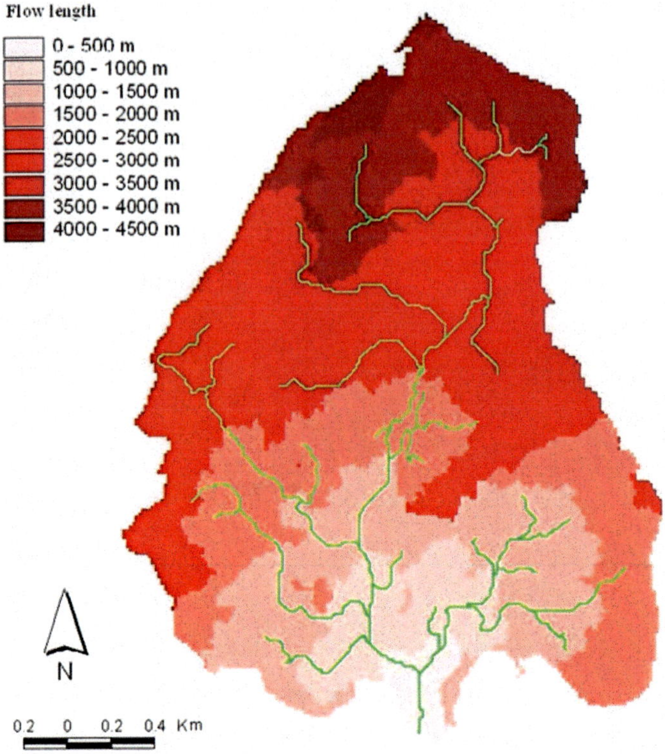

Fig. 3.5 Flow lengths in the Shangshe catchment

estimate of the concentration time is obtained using Irish's formula (Jain and Singh, 2000):

$$t_c = \sum_{i=1}^{n} t_{ci} = \sum_{i=1}^{n} 0.06628 L_i^{0.77} H_i^{-0.305} \tag{3.2}$$

where

t_{ci} is the time of travel in hours of each catchment;
t_c is the concentration time in h;
L_i is the length of main stream in km;
H_i is the average slope of the main streams (m km^{-1}); and
i is the number of catchments from the i point to the river outlet.

Substituting the values of L_i and H_i in Eq. (3.2), the time of concentration t_c was computed. The time–area diagram was developed (Fig. 3.6). The areal distribution of various land types within different concentration time was calculated by geoprocessing Figs. 3.2 and 3.6 under ArcView (shown in Fig. 3.7).

The formula used to calculate the total *ISSD* (instantaneous suspended sediment discharge, kg s^{-1}) at t time at the river outlet is Eq. (3.3) and the formula used to calculate the *SISSD* at t time at the river outlet is Eq. (3.4).

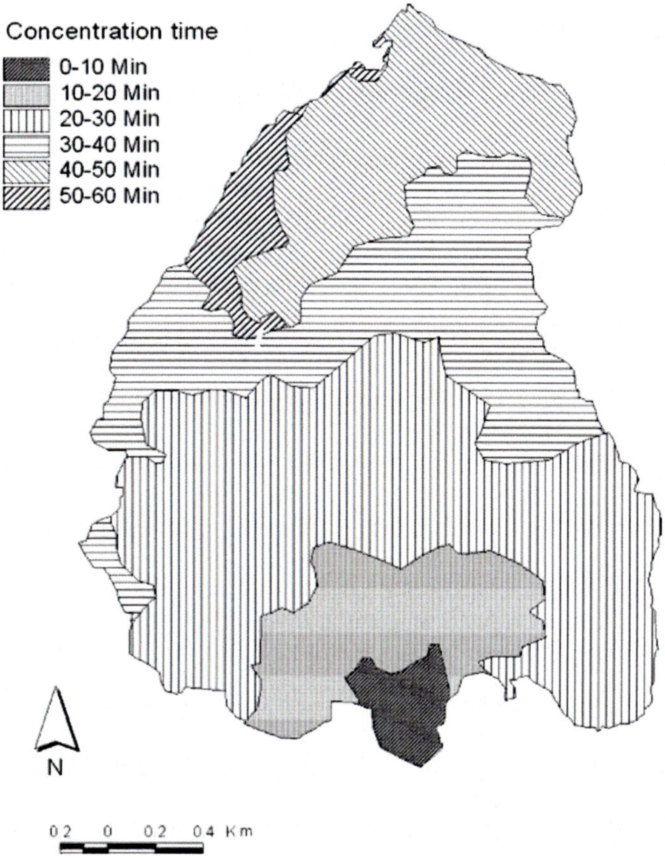

Fig. 3.6 Concentration time in the Shangshe catchment

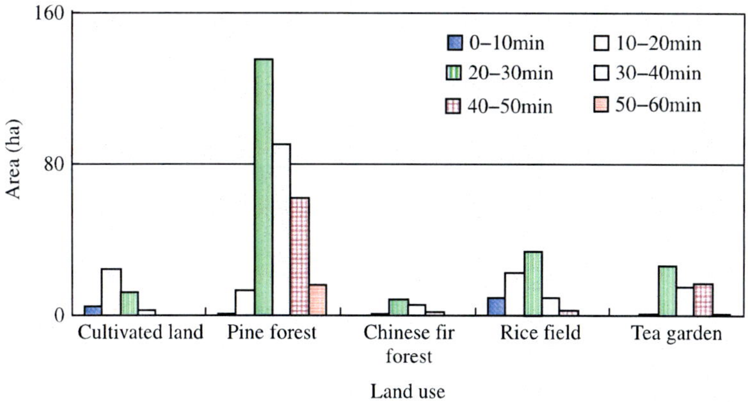

Fig. 3.7 Areas within different concentration times for various kinds of land use in the Shangshe catchment

There are two assumptions concerning the formulae:

(1) Each observed *SISSD* graph is assumed to be a unit, which keeps its shape during travel to the outlet of the catchment. Neither sedimentation nor gully erosion is considered.
(2) The geomorphologies of the four monitored sites are thought of as representatives of every corresponding kind of land use within the Shangshe catchment. Each observed *SISSD* graph can thus be applied to other places, as the rainfall-sediment discharge response for the same rainfall event is assumed the same if the land use type is the same, despite the fact that coverage, slope, and soil characteristics may be different from those of the four kinds of monitored sites:

$$Q_{\mathrm{sdt}} = \sum_{i=1}^{n} Q_{\mathrm{sdt}i} = \sum_{i=1}^{n}\sum_{j=1}^{m} S_{ij} Q_{\mathrm{SISSD}tij} \tag{3.3}$$

$$Q_{\mathrm{SISSD}t} = Q_{\mathrm{sdt}} / \sum_{i=1}^{n} S_i \tag{3.4}$$

where

Q_{sdt} is the total *ISSD* at the river outlet at *t* time (kg s^{-1});

$Q_{\mathrm{sdt}i}$ is the *ISSD* at *t* time at the river outlet from the *i*th land use of the catchment (kg s^{-1});

i is the ID number of the land use;

j is the ID number of concentration time;

S_{ij} is the area of the *i*th land use within the *j*th interval of concentration time (km^2);

$Q_{\mathrm{SISSD}tij}$ is the *SISSD* at the river outlet at the *t* time from the *i*th land use within the *j*th interval of concentration time calculated from observed *SISSD* graph of the *i*th land use (kg km^{-2} s^{-1}), $Q_{\mathrm{SISSD}tij} = Q_{\mathrm{OSISSD}_{ti+tj}}$, where t_i is the time it takes for the observed *SISSD* graph of the *i*th land use to occur after a rainfall, t_j is the concentration time from the *j*th interval of concentration time in the example (Fig. 3.6), t_i=0; and

$Q_{\mathrm{SISSD}t}$ is the *SISSD* at the river outlet at the *t* time (kg km^{-2} s^{-1}).

3.4 Results and Discussion

3.4.1 The Results of Calculated SISSD Graph Compared with Observed Ones

The calculated *SISSD* graphs compared with observed graphs of 18 events in the year 2000 are shown in Figs. 3.8–3.25. The results show that for the big and medium-sized events in summer, the observed *SISSD* graphs compared quite well with observed ones (Figs. 3.9, 3.10, 3.12, 3.14, 3.15, 3.16, 3.18, and 3.20). But for small events (Figs. 3.8, 3.11, 3.13, 3.17, 3.19, 3.21, 3.23, 3.24, and 3.25) and one medium-sized event (Fig. 3.22), they did not agree well.

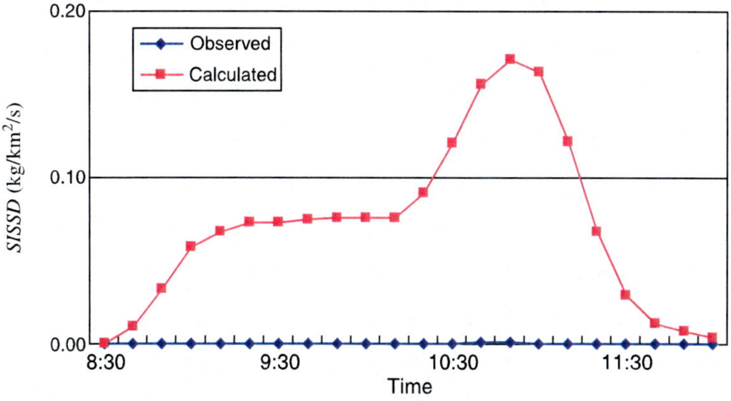

Fig. 3.8 Comparison of the observed *SISSD* graph with the one calculated at the river outlet on 9 May 2000

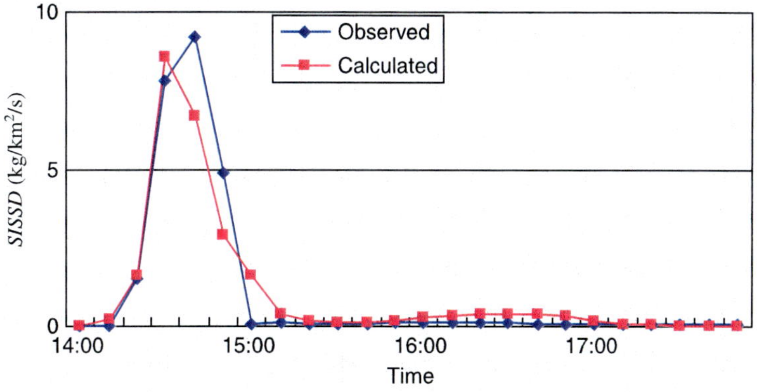

Fig. 3.9 Comparison of the observed *SISSD* graph with the one calculated at the river outlet on 15 May 2000

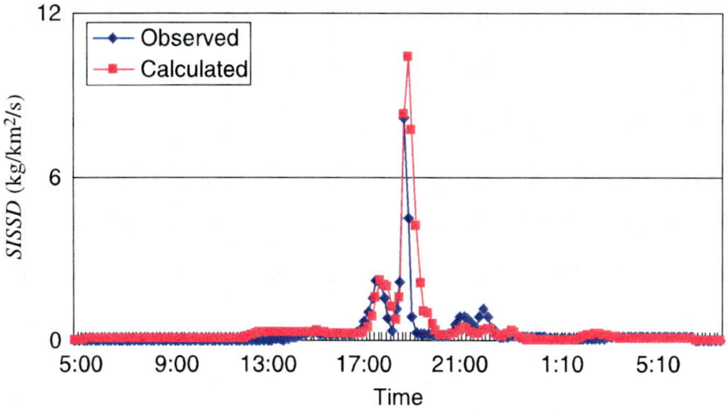

Fig. 3.10 Comparison of the observed *SISSD* graph with the one calculated at the river outlet on 25 and 26 May 2000

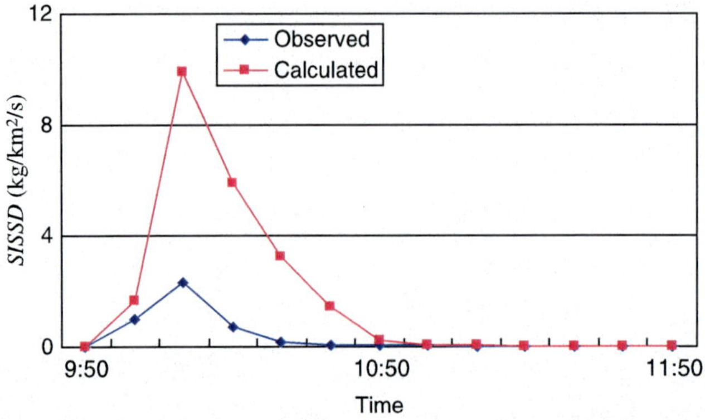

Fig. 3.11 Comparison of the observed *SISSD* graph with the one calculated at the river outlet on 3 June 2000

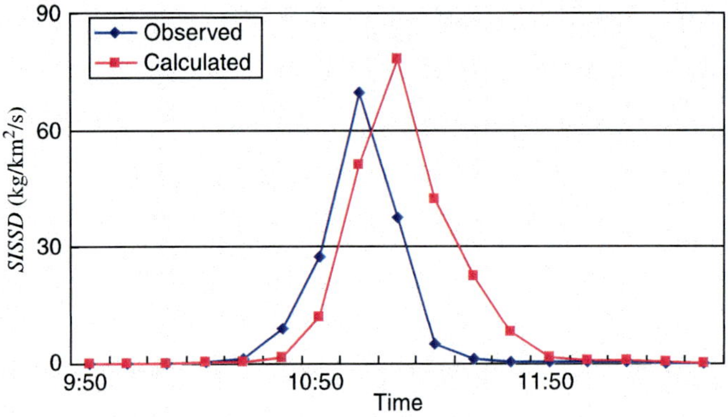

Fig. 3.12 Comparison of the observed *SISSD* graph with the one calculated at the river outlet on 22 June 2000

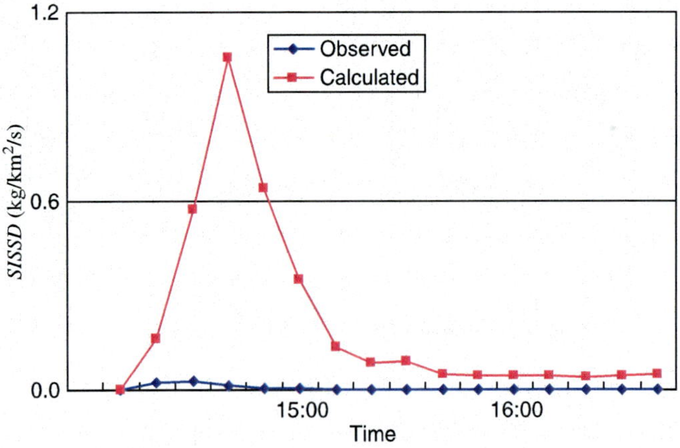

Fig. 3.13 Comparison of the observed *SISSD* graph with the one calculated at the river outlet on 6 August 2000

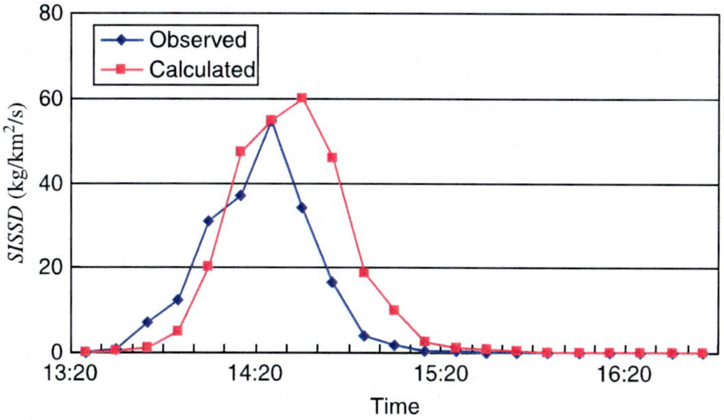

Fig. 3.14 Comparison of the observed *SISSD* graph with the one calculated at the river outlet on 8 August 2000

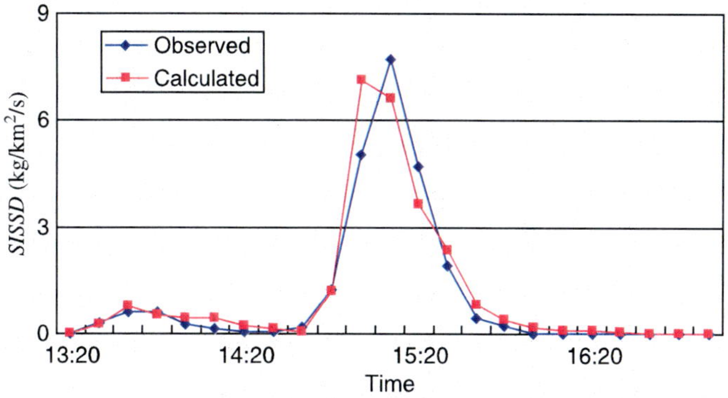

Fig. 3.15 Comparison of the observed *SISSD* graph with the one calculated at the river outlet on 13 August 2000

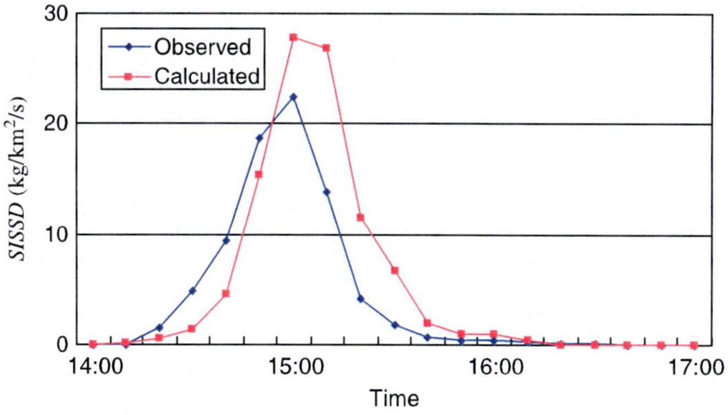

Fig. 3.16 Comparison of the observed *SISSD* graph with the one calculated at the river outlet on 14 August 2000

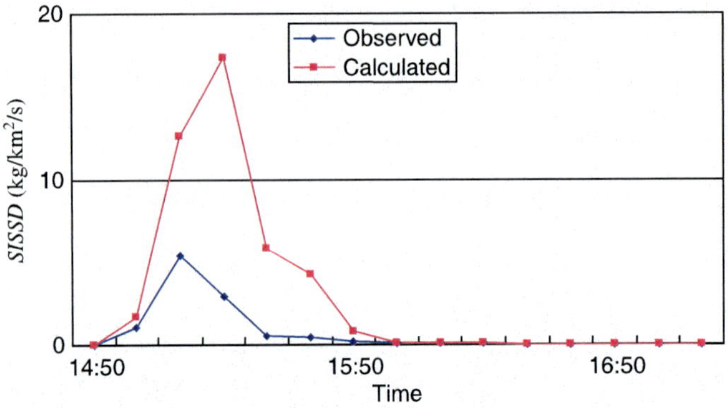

Fig. 3.17 Comparison of the observed *SISSD* graph with the one calculated at the river outlet on 15 August 2000

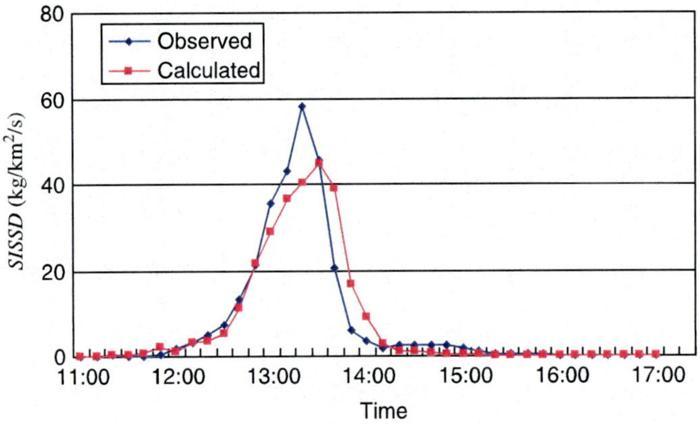

Fig. 3.18 Comparison of the observed *SISSD* graph with the one calculated at the river outlet on 16 August 2000

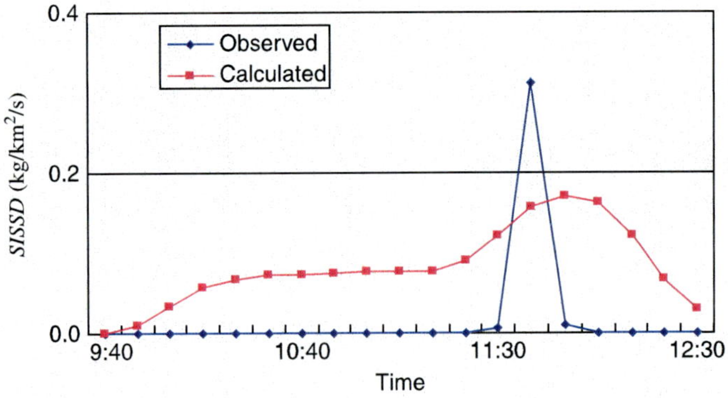

Fig. 3.19 Comparison of the observed *SISSD* graph with the one calculated at the river outlet on 17 August 2000

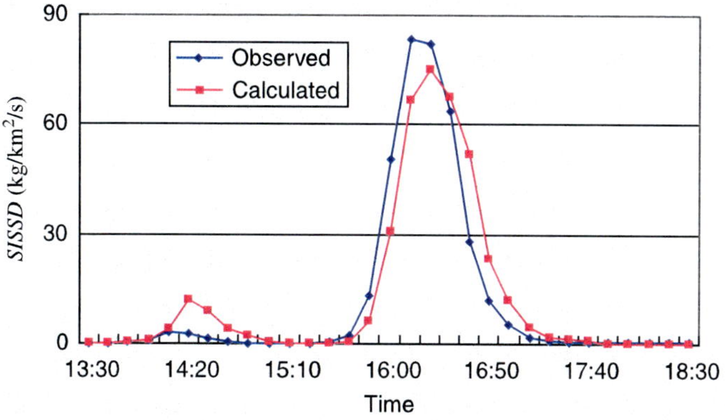

Fig. 3.20 Comparison of the observed *SISSD* graph with the one calculated at the river outlet on 18 August 2000

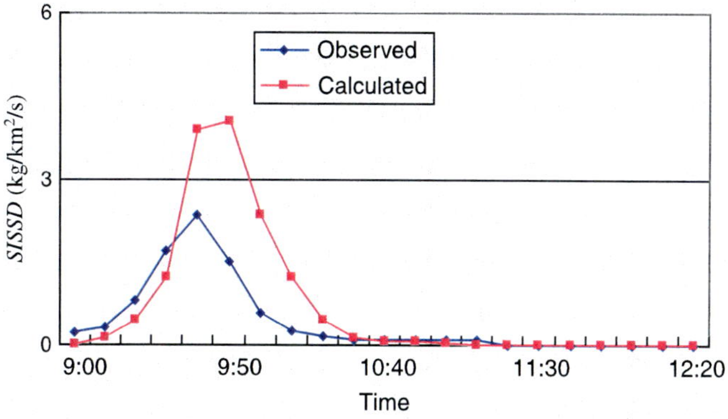

Fig. 3.21 Comparison of the observed *SISSD* graph with the one calculated at the river outlet on 19 August 2000

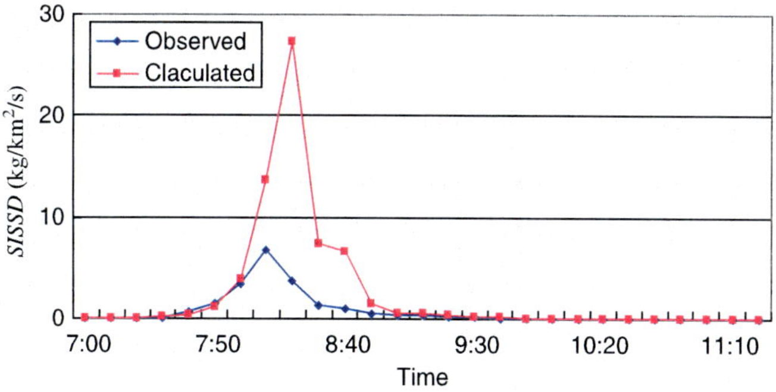

Fig. 3.22 Comparison of the observed *SISSD* graph with the one calculated at the river outlet on 5 September 2000

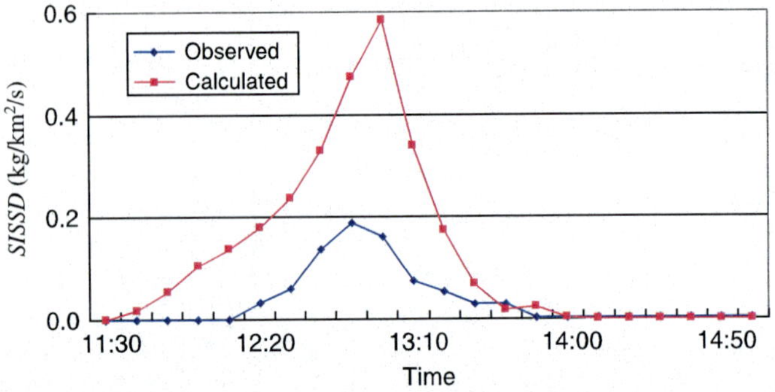

Fig. 3.23 Comparison of the observed *SISSD* graph with the one calculated at the river outlet on 25 September 2000

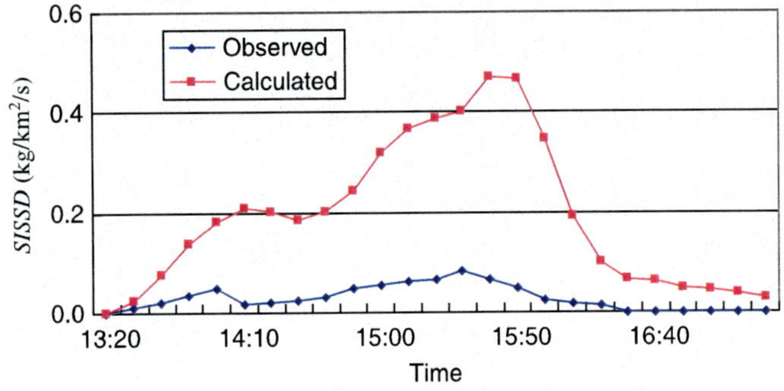

Fig. 3.24 Comparison of the observed *SISSD* graph with the one calculated at the river outlet on 2 October 2000

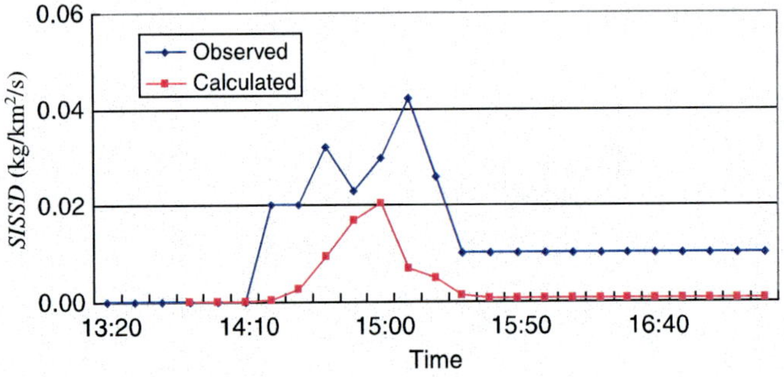

Fig. 3.25 Comparison of the observed *SISSD* graph with the one calculated at the river outlet on 5 October 2000

The *SSD* for each single event in the year 2000 was calculated using Eq. (2.11). Table 3.1 shows that the 18 events were divided into three groups. (1) Small events are those for which the ratio of *SSD* for single events to annual *SSD* (%) is less than 1%; (2) medium events are those for which the ratio of *SSD* for single events to

Table 3.1 Classification of rainfall–runoff events in the year 2000

	Ratio of *SSD* for single events to annual *SSD* (%)	Number	Ratio of *SSD* for three groups to annual *SSD* (%)
Large events	>5	5 (Figs. 3.12, 3.14, 3.16, 3.18, and 3.20)	89.00
Medium events	1–5	4 (Figs. 3.9, 3.10, 3.15, and 3.22)	9.07
Small events	0–1	9 (Figs. 3.8, 3.11, 3.13, 3.17, 3.19, 3.21, 3.23, 3.24, and 3.25)	1.93

annual *SSD* (%) is between 1 and 5%; and (3) large events are those for which the ratio of *SSD* for single events to annual *SSD* (%) is more than 5%. The ratio of *SSD* for the three groups (small, medium, and large) to annual *SSD* is 1.93, 9.07, and 89%, respectively.

3.4.2 Discussion

From the combination of the observed *SISSD* graph at the various monitored sites with GIS and a hydrology model, a GOIUG model was developed to simulate the *SISSD* graph at the river outlet of the Shangshe catchment. For big events and most medium-sized events, the calculated *SISSD* graph compared well with the observed ones. But for small events, they did not agree. Because the ratio of *SSD* for small events to the annual total *SSD* is only 1.93%, despite the fact that model and observed small events did not agree, the small events can be neglected in this study.

This method tried to establish a linkage between sediment export and intra-basin processes according to land use in the Shangshe catchment. However, this model used only the observed *SISSD* graphs for various kinds of land use for each rainfall event as the corresponding basic unit graphs of rainfall-sediment discharge response events to simulate the sediment discharge graphs at the river outlet. Neither sedimentation nor gully erosion was considered during the traveling of the unit graphs. So this model is specifically suitable for catchments with observational data of representative kinds of land use. In this research, the observed results were applied to other corresponding areas for the same rainfall event if the land use was the same. The differences of the geomorphology as well as the soil and the coverage among different catchments of the same land use were neglected; but in fact, these are very important factors. Therefore, more observational data are needed to make empirical rainfall–runoff–*SISSD* response curves for the prediction of *SISSD*. So in the future, development of GOIUG model will aim to combine the observed graphs of various land use types with the USLE (Wischmeier and Smith, 1978), as well as to make empirical *SSC* curves for more extensive application of this model.

References

Crawford NH, Linsley RK (1966) Digital simulation in hydrology: Stanford watershed model IV, Technical Report No. 39. Civil Eng, Stanford University

De Roo APJ (1993) Modeling surface runoff and erosion in catchments using geographical information systems: validity and application of the "ANSWERS 'model in two catchments in the loess area of South Limburg (UK)'". Neth Geogr Stud 157:304

De Roo APJ (1998) Modeling runoff and sediment transport in catchments using GIS. Hydrol Process 12(6):905–922

De Roo APJ, Wesseling CG, Ritsema CJ (1996) LISEM: a single-event physically-based hydrological and soil erosion model for drainage basins: Theory input and output. Hydrol Process 10:1107–1118

Edmonds RL (1992) The Sanxiadz (Three Gorges) Project: the environmental argument surrounding China's super dam. Global Ecol Biogeogr 2:105–125

Gregory KJ, Walling DE (1973) Drainage basin form and process, a geomorphological approach. Edward Arnold, London

Jain SK, Singh RD (2000) Design flood estimation using GIS supported GIUH approach. Water Resour Manage 14:369–376

Jean S (2004) Change in river suspended sediment concentration in central Japan in response to late 20th century human activities. Catena 55(4):231–254

Lee KT (1998) Generating design graphs by DEM assisted geomorphic runoff simulation: a case study. J Am Water Resour Ass 34(2):375–384

Mario A, Lorenzo M (2000) Suspended sediment load during floods in a small stream of the Dolomites (Northeastern Italy). Catena 39:267–282

Milliman JD, Syvitski JPM (1992) Geomorphic/tectonic control of sediment discharge to the ocean: the importance of small mountainous rivers. J Geol 100:525–544

Moore ID, Grayson RB (1991) Digital terrain modeling: a review of hydrological, geomorphological, and biological applications. Hydrol Process 5:3–30

Pascal S, Laura B (1998) Application of a GIS-Based distributed hydrology model for prediction of forest harvest effects on peak stream flow in the Pacific Northwest. Hydrol Process 12 (6): 889–904

Rode M, Lindenschmidt KE (2000) Distributed sediment and phosphorus transport modeling on a medium sized catchment in Central Germany. Phys Chem Earth (B) 26:635–640

Rodriguez-Iturbe I, Valdez, JB (1979) The geomorphologic structure of hydrology response. Water Resour Res 15(6):1409–1402

Rudra PR, Dickinson WT, Wall JG (1995) Problems regarding the use of soil erosion models. Modeling soil erosion by water. In: Boardman J, Favis-Mortlock D (eds) NATO ASI Series I. Global Environ Chang 55:175–189

Su B, Sokazama M, Masaki S (2003) Development of a distributed hydrological model and its application to soil erosion simulation in a forested catchment during storm period. Hydrol Process 17:2811–2823

Summfield H (1994) Natural controls of fluvial denudation rates in major world drainage basin. J Geophys Res 99:1371–1383

Walling DE (1996) Use of fallout radionuclide measurements in sediment budget investigations. Geomorphologies: Relief, Process Environ 2:41–54

Wischmeier WH, Smith, DD (1978) Predicting rainfall erosion losses, a guide to conservation planning, agricultural handbook 537. US Department of Agriculture, Washington, DC

Xiang ZA, Zhou GY (1986) An analysis of sediment transport in the Changjiang River. Sediment Inf 22:8–13 (in Chinese)

Yang SI, Zhao QY, Belkin IM (2002) Temporal variation in the sediment load of the Yangtze River and the influences of human activities. J Hydrol 263:56–71

Chapter 4
GIS-Based ER-USLE Model to Predict Soil Loss in Cultivated Land

Abstract In the Dabie Mountains of the lower Yangtze basin, China, soil loss reduction has been deemed as important as flood control after the 1998 flood, which caused great damage to the area and its residents. For soil loss control in the Dabie Mountains, China, the first step is the determination of appropriate strategies. An urgent need exists for soil loss models that can provide sufficient information for developing strategies to control soil loss using field observation data over several years. In this study, field observations of water and sediment discharges were carried out at the micro-plot scale and the USLE (universal soil loss equation)-plot scale in the Shangshe catchment of the Dabie Mountains in China. Through analyses of field observation data in the Shangshe catchment during the period 1999–2007, the USLE model was applied to predict both annual soil loss and that for single events based on GIS in a sub-catchment of cultivated land. Model calibration showed that by using an effective rain erosivity factor (R_e) instead of R (rain erosivity) in the RUSLE, and by portraying the interaction among seasonal precipitation, seasonal crop coverage (C_s), and practice of discrete-event practices by human beings (P_s) in the USLE, an ER-USLE model was developed and model accuracy improved. With these measures, the ER-USLE model can be utilized to predict annual soil loss based on single events from field observation data over a few years. Comparison between predicted results and observed results of the field observation data of soil loss from 2000 to 2007 in the sub-catchment of cultivated land scale with USLE and ER-USLE models shows that the R^2 of ER-USLE (0.82) is much higher than that (0.46) of the USLE. With ER-USLE and GIS, strategies for soil loss control are possible based on seasonal soil loss information with field observation data over a few years.

4.1 Introduction

The universal soil loss equation (USLE), an empirical model for predicting the average annual soil loss caused by rainfall, was originally developed by Wischmeier and Smith (1965). The USLE has been widely applied at the watershed scale on the basis of an aggregated approach (Williams and Berndt, 1972, 1977; Wilson, 1986; Griffin et al., 1988; Dickinson and Collins, 1998). GIS development, which reduces the time needed for analyses, has allowed for the application of USLE with a spatially distributed approach. It allows one to estimate average annual soil losses for given

J.C. Zhang et al., *Theory and Practice of Soil Loss Control in Eastern China*,
DOI 10.1007/978-1-4419-9679-4_4,

natural and anthropogenic conditions. It was created as a support to soil conservation planning at the field scale. RUSLE (revised USLE; Renard et al., 1997) was developed in order to include additional data gained since the mid-1970s (Kinnell, 1998). MUSLE (modified USLE; Williams, 1972) was developed for estimating the sediment load produced by each storm. MUSLE incorporates the volume of runoff rather than the rainfall erosivity. But this change from rainfall erosivity to volume of runoff was made without regard to fundamental mathematical principles (Kinnell, 2004).

In the Dabie Mountains of the lower Yangtze basin, China, soil loss reduction has been deemed to be as important as flood control after the 1998 flood (Zhang et al., 1999), which caused great damage to the area and its residents. Accurate and correct prediction of soil losses for single rainfall–runoff events, as well as of the annual average soil loss, is important for the development of soil loss control strategies. Application of the USLE and RUSLE model can provide average annual soil loss information only when it is provided with observation data for scores of years. With the development of the economy in China, controlling environmental problems has become so important that more information should be provided for making soil loss control strategies from field observation data over periods of only a few years rather than scores of years. The USLE is based on a clean-tilled fallow plot with a length of 22.1 m, width of 5 m, and a slope of 9% (Wischmeier and Smith, 1965). Usually a plot of 22.1 m with a width of 5 m is referred to as a USLE plot and a plot less than 2 m long and 2 m wide is called as micro-plot.

This study investigated soil loss at the micro-plot level and the USLE plot level to develop a soil loss model suitable for predicting annual soil loss based on single rainfall–runoff events in the Dabie Mountains, China, using field observation data collected over only a few years.

4.2 Materials and Methods

4.2.1 Study Area

This study was conducted in the Shangshe catchment (34°32′20″N, 116°50′20″E,) of the Dabie Mountains in Yuexi prefecture of Anhui Province, China. Details of the Shangshe catchment were reported by Zhuang et al. (2004). In the Shangshe catchment, a sub-catchment of cultivated land of 7400 m^2 was selected for field observations. Terraces were built in the sub-catchment. The width of terraces was 2–5 m in the lower slope and 1–2 m in the middle and upper slopes. The crops were wheat and oilseed rape from winter to summer, and corn from summer to autumn. Soil depth is about 16 cm and the soil is sandy loam derived from weathering granite. The main human activities in the cultivated land are harvesting of wheat in June, plowing land in early July for corn planting, and weeding and fertilization in August.

4.2.2 *Field Observations at the USLE-Plot Scale and the Micro-plot Scale*

The method of using micro-plots at a meter scale was applied in the sub-catchment of cultivated land for the year 2000. Ten micro-plots of 2 m × 1 m on slopes of 1°, 3°, 5°, 6°, 8°, 11°, 17°, 20°, 25°, and 35° were set up, whereas a USLE plot of 22.1 m × 5 m was set up on a slope of 23° with level terraces within the sub-catchment and another USLE plot of the same size was on a slope of 25° with up-and-down tillage 200 m away from the sub-catchment cultivated land (Fig. 4.1). Despite the fact that terraces were built in the sub-catchment on the lower slope, there were a few scattered narrow slopes without terraces and they were selected for the experimental placement of micro-plots of 17°, 20°, 25°, and 35° slopes. The other micro-plots were set up between the terraces. For each micro-plot, a trench was installed at the lower boundary of the plot and one container was used to collect water and sediment discharge, whereas for the USLE plots, one collecting tank of 1.5 m × 1.2 m × 1.2 m and one collecting tank of 1 m × 1 m × 1 m were built. Sediment and surface runoff from the plot entered the first small collecting tank and then entered the second one when the height of water in the small tank was higher than 50 cm. The volume of surface runoff was calculated by measuring the height of the water in the first and second collecting tanks. In each micro-plot and USLE plot, the water and sediment discharges were mixed and a 200–500 ml sample was taken to the laboratory where the organic matter, such as grass and leaves, was picked out and sediment was filtered, dried in an oven at 110°C, and weighed. For a few extreme events, when the two tanks were full, the observer measured the runoff in the two tanks and took a sample from the tanks. Then the runoff in the two tanks was discharged to continue the field observations. For each rainfall–runoff event, the runoff volume and the soil sediment loss from the micro-plots and the USLE plot were calculated. Field observations at the micro-plot scale were carried out during

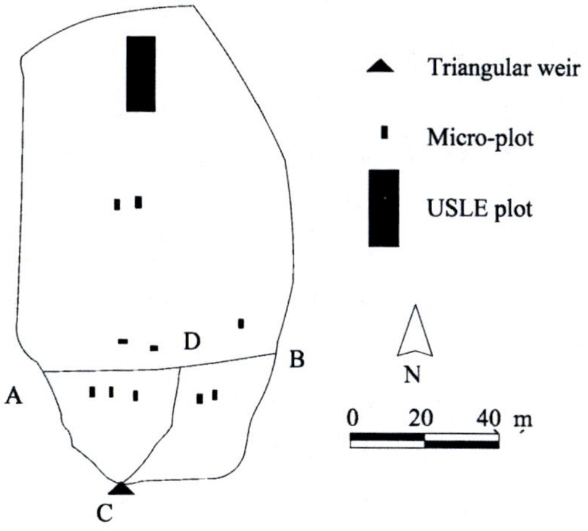

Fig. 4.1 Measuring devices in the sub-catchment of cultivated land

the year 2000 and field observation at the USLE plot scale was carried out from 2000 to 2003. Tipping bucket rain gauge data for the years 1999 and 2000 were used and manual rain gauge data of precipitation for years 2001 and 2002 were used.

4.2.3 Field Observations at the Sub-Catchment Scale

A 45° V-notch, sharp-crested weir, equipped with a float-type water level recorder, was built at the outlet of the sub-catchment of cultivated land (Fig. 4.1). At 8 a.m. each day, the recorder paper of a water gauge was changed manually; the water level at the weir was also measured manually for use in correcting any error of the water level in the recorder paper. Details of water runoff and suspended sediment concentration observations, as well as runoff and suspended sediment discharge calculations on the sub-catchment scale, are described in Zhang et al. (2004). The sediment at the weir was measured and carried out for each rainfall–runoff event. The sediment discharge (SD, t) was transformed into specific sediment discharge (SSD, t km^{-2}). Field observation data of SSD for single events in 2000 were used in this study. Field observation data of precipitation in years 2000–2002 were used in this study.

4.2.4 Precipitation Observation

A manual rain gauge and an autographic rain gauge were set up to survey precipitation and rainfall intensity 400 m away from the weir at the monitored site of Chinese fir forest. The recorder paper of the autographic rain gauge was changed daily at 8 a.m. For every rainfall event, precipitation was read from the recorder paper for the automatic rain gauge and compared with data from the manual rain gauge. Field observation data for precipitation from 1999 to 2002 were used in this study.

4.2.5 Proposal for Use of ER-USLE Model for Annual Soil Loss Prediction Based on Single Events

The USLE, an empirical model for predicting the average annual soil loss caused by rainfall, was originally developed by Wischmeier and Smith (1965). This is represented by the following equation:

$$A = RLSKPC \tag{4.1}$$

where

A is the annual average soil loss (g m^{-2} or t km^{-2});
R is the effective rain erosivity factor (J mm m^{-2} h^{-1});
L is the slope length factor;
S is the slope factor;

P is the support-practice factor, which accounts for control practices that reduce the erosion potential of the runoff by their influence on drainage patterns, runoff concentration, runoff velocity, and hydraulic forces exerted by runoff on soil. *L*, *S*, and *P* are dimensionless;

K is the soil erodibility factor (g J^{-1} mm^{-1} h^{-1}).

Application of the USLE model can provide average annual soil loss information only when it is calibrated with observation data for many years. It takes little account of the seasonal variation of coverage or of discrete-event practices (see below) of humans in the cultivated land. With the growth of the economy, the control of environmental problems has become so important that more information should be provided for developing soil loss control strategies based on field observation data over periods of only a few years, rather than requiring several years.

Human activity in the cultivated land includes two aspects that are relevant to its effect on soil loss: one is the support-practice factor (*P*) of the USLE, which represents the effect of the conservation support-practice factor on long-term soil loss. The *P* factor is the soil loss ratio that occurs when a specific support practice is used to the corresponding soil loss that occurs with up-and-down slope tillage (Renard et al., 1997). This factor affects soil loss over the long term. Usually this factor decreases soil loss. On the other hand, to predict soil loss for single rainfall–runoff events, the application of discrete-event practices in the cultivated land (P_s), such as seed planting and cleaning the grass from the ground in cultivated land (which affect soil losses in different ways), must be determined. The P_s factor (for discrete-event practices) is the ratio of soil loss that occurs associated with a specific practice to the corresponding soil loss without that practice. Usually it increases soil loss.

Similarly, the coverage factor is divided into two sub-factors: land use sub-factor (C_l) and seasonal coverage sub-factor (C_s). The parameter C_l is the ratio of observed soil loss to R_eLSKP for a certain type of land use. In the Dabie Mountains, the distribution of precipitation in a year is uneven. To predict soil loss for single rainfall–runoff events, the seasonal coverage factor in the cultivated land (C_s) must be determined. The C_s factor (seasonal coverage) is the ratio of soil loss with a specific crop coverage to the corresponding soil loss predicted from USLE in a certain season.

Taking account of the P_s, C_l, and C_s factors into USLE model, Eq. (4.2) is obtained for soil loss prediction based on single events:

$$A_e = R_eLSKPP_sC_lC_s \tag{4.2}$$

where

A_e is the *SSD* (specific sediment discharge) for single events (t km^{-2} or g m^{-2}) (at the micro-plot and USLE-plot scale; the unit is (g m^{-2}), while at the sub-catchment and the catchment scale, the unit is transferred into (t km^{-2}) in this study);

R_e is the effective rain erosivity factor (J mm m^{-2} h^{-1});
L is the slope length factor and S is the slope factor;
P_s is the discrete-event practice factor in cultivated land;
P is the support-practice factor which affects soil loss over a long period;
K is the soil erodibility factor (g J^{-1} mm^{-1} h);
C_l is the factor of land use;
C_s is the seasonal coverage factor.

Equation (4.3) is obtained for annual or seasonal loss prediction based on summation over single events:

$$\sum_{i}^{n} A_{ei} = \sum_{i=1}^{n} R_{e_i}\ LSKPP_{si}C_l C_{si} \tag{4.3}$$

where

$\sum_{i}^{n} A_{ei}$ is the annual or seasonal SSD (t km^{-2} or g m^{-2});
n is the number of the precipitation events and i represents the ith event of the year; the other factors are the same as those in Eq. (4.2).

Because this model is developed from the USLE with an effective R_e factor, it is called the ER-USLE model.

4.3 Calculation of the Factors Used in the ER-USLE Model

4.3.1 R_e Factor

Yang and Guo (1994), using observed data from normal runoff plots in cultivated land, analyzed the relationship between the R values calculated at different time intervals (5 min to 1 h) and the amount of soil loss for every event. They reported that the soil loss at the USLE-plot scale correlated excellently with an R value calculated at 10-min intervals in northern China. Therefore, their method was applied in this study. The equation for R is (Wischmeier and Smith, 1965)

$$R = \sum_{i}^{n} E_i I_{30} \tag{4.4}$$

where

R is the rainfall–runoff erosivity factor due to precipitation (J mm m^{-2} h^{-1});
E_i is the rain energy in one storm during the ith 10-min interval;
I_{30} is the maximum continuous rain intensity during a 30-min period in a storm (mm h^{-1}).

The Committee of the Yellow River of China has also computed a regression equation that incorporates rain energy and intensity (Zhang and Hu, 1996):

$$E = 210.3 + 89\log_{10} I \tag{4.5}$$

where

E is the rain energy during one time period (J m^{-2} cm^{-1}) and
I is the rain intensity (cm h^{-1}).

This regression equation was applied to calculate E in this study.

In general, the erosion models have difficulty predicting small-scale events, due to the large natural variation of factors related to soil loss data (Neering, 1998). For many small rainfall events, there is no runoff at all; for most rainfall–runoff events, rain loss (runoff) varies greatly. The reason for this phenomenon may be attributed to the initial rain loss, which varies with the interval between the event in question and the preceding precipitation event. For example, Minami et al. (2002) reported that the initial loss of rainfall to the surface flow runoff increased from 4.5 to 10 mm when the rain interval days increased from 1 to 8 days. Using the R factor without considering the variation in rain runoff loss will over-predict soil loss for small-scale events. Therefore, a modified effective rainfall erosivity was proposed to smooth the error of rain runoff loss on soil loss in the USLE application in China.

In 1999 and 2000 in the Shangshe catchment, 28 and 18 rainfall–runoff events, respectively, occurred in the sub-catchments of Chinese fir. Because the runoff observations were carried out at the sub-catchment of Chinese fir, by comparing the hydrograph of runoff with the hyetograph of precipitation, the loss of rainfall could be determined. Equation (4.6) was applied to calculate the R_e in this study:

$$R_e = (E - E_l)I_{30} \tag{4.6}$$

where

R_e is the modified R factor (J mm m^{-2} h^{-1});
E is the rain energy of one rainfall–runoff event (J m^{-2});
E_l is the rain loss energy that takes place up to the point of the occurrence of rainfall–runoff during the same rainfall–runoff event (J m^{-2}).

4.3.2 LS Factor

4.3.2.1 *LS* Factor at the Micro-plot Scale

Analyses of field observation data of 18 rainfall–runoff events at the micro-plot scale showed that *WD* (water discharge, mm) and *SSD* increased with increasing slope, despite the fact that the relationship between slopes and *SSD* at the micro-plot scale varied among events. The 16 August 2000 event is shown as an example (Fig. 4.2).

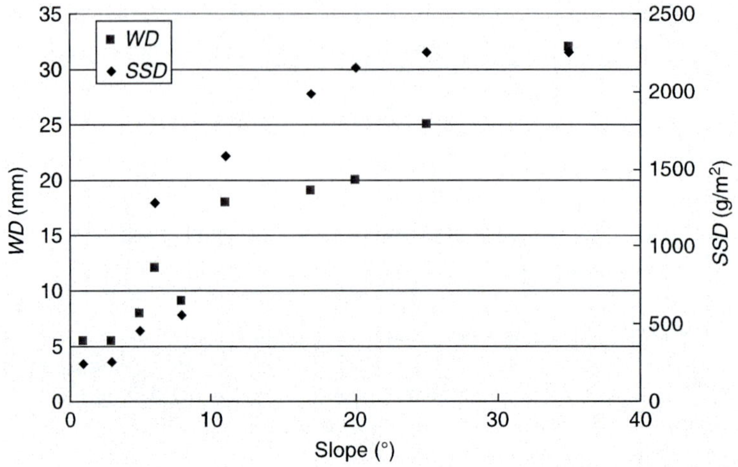

Fig. 4.2 Relationship among *WD, SSD,* and slope at the micro-plot scale on 16 August 2000

The value of *LS* for each micro-plot was calculated according to the following formula (Moore and Burch, 1986):

$$LS = (\lambda/22.1)^{0.6}\left(\frac{\sin\theta}{0.0896}\right)^{1.3} \tag{4.7}$$

where

λ is the length of slope in m
θ is the slope in degrees.

4.3.3 LS Factor at the Sub-catchment Scale

The slope was derived from the digital elevation map (DEM) of the sub-catchment of cultivated land at the 1-m×1-m grid scale of resolution (Fig. 4.3). In both the USLE and RUSLE models, slope length is defined as the horizontal distance from the origin of overland flow to the point where either the slope gradient decreases to a point at which deposition begins or runoff becomes concentrated in a defined channel (Wischmeier and Smith, 1965; Renard et al., 1997). In the present study, the sub-catchment was divided into three segments (Fig. 4.1) by a terrace (line AB) and a stream (line CD): the upslope segment and two side segments along the stream. The average slope length in the upslope segment is the ratio of its area to the length of line AB, and the average slope lengths of the two side segments along the stream are the ratios of their areas to the length of stream CD.

The *LS* was calculated according to Eq. (4.7) (Moore and Burch, 1986) and is shown across the sub-catchment in Fig. 4.4.

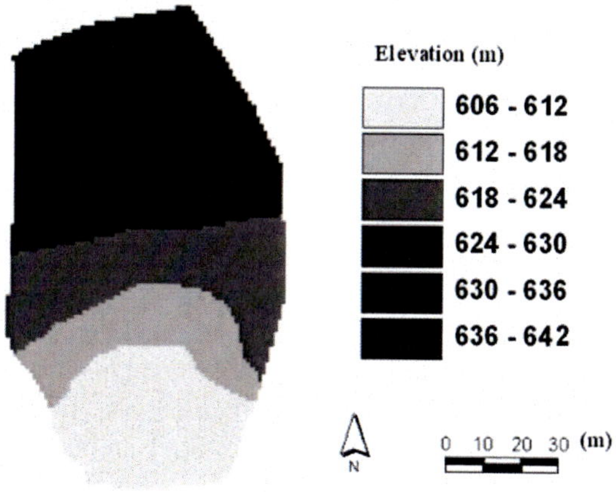

Fig. 4.3 Digital elevation map (DEM) of the sub-catchment of cultivated land

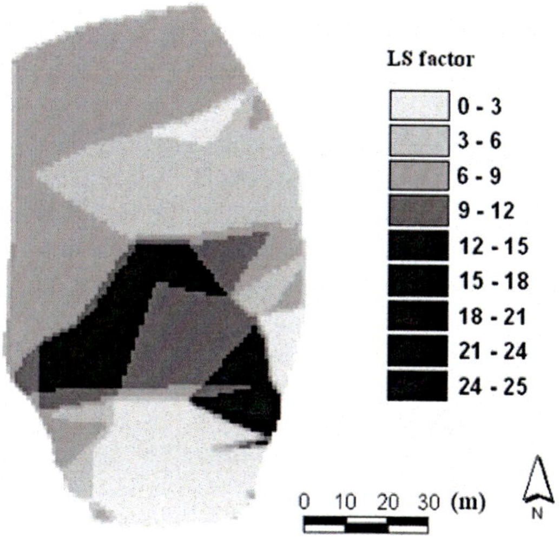

Fig. 4.4 *LS* factor on the sub-catchment of cultivated land

4.3.4 K Factor

K is calculated with Eq. (4.8) (Williams et al., 1984):

$$K = \left\{0.2 + \mathrm{e}^{[-0.0256 S_1 (1 - S_1/100)]}\right\} \left[S_2/(S_3 + S_2)\right]^{.3}$$
$$\left\{1 - 0.25C/\left[C + \mathrm{e}^{(3.72 - 2.95C)}\right]\right\} \left\{1 - 0.7n/\left[n + \mathrm{e}^{(-5.51 + 22.95n)}\right]\right\} \quad (4.8)$$

where

S_1 is the percentage of sand grains;
S_2 is the percentage of silt;

S_3 is the percentage of clay;
C is the percentage of organic matter and where $n = 1 - S_1/100$.

4.3.5 C_l Factor

The C_l factor reflects the effect of land use or crop types on soil erosion rates. The vegetation cover and management factor (C) in the universal soil loss equation (USLE) is an index to incorporate this effect, which has been studied intensively and used widely in the USA. In China, research on the effect of crop and cultivation style on soil loss is still insufficient, although the cultivation of land is a key factor in soil loss (Zhang et al., 2002). In some research, C has been estimated by establishing a linear regression between coverage and the C value (Bu et al., 1993; Jiang et al., 1996; Jin et al., 1992; Ma, 1989). In this research, even for the same crop there were great differences among the C values, because the investigators failed to consider the seasonal distribution of precipitation. Zhang et al. (2002) suggested that the cropland C factor research should be studied further, and the research methodology should be unified across China to obtain reliable C values for soil loss prediction and conservation planning. The C factor depends not only upon the coverage and rotation of crops and management but also upon the precipitation distribution over different growing seasons, and especially upon the erosivity of precipitation during the period when the soil erosion control function is the least effective (Wischmeier and Smith, 1978). To take this into account in the ER-USLE model, the C factor is divided into two sub-factors: the land use sub-factor (C_l) and the seasonal coverage sub-factor (C_s). The C_l factor reflects the effect of land use or crop types on soil erosion rates. C_l is thought of as a constant for one type of land use with certain crops in a year. In one type of land use, it is assumed that values of SSD in the micro-plots show a linear regression relationship with R_eLSPK. The constant parameter of the linear regression model between SSD and R_eLSPK is defined as C_l. This method is used to determine the C_l factor. The seasonal coverage sub-factor (C_s) is discussed in Section 4.3.7.

Using the field observation data of 18 rainfall–runoff events, the relationship between R_eLSPK and SSD at the micro-plot scale was analyzed. Figure 4.5 shows the relationship between R_eLSPK and SSD at the micro-plot scale. A linear regression was performed, showing that $C_l = 0.17$. However, the data in Fig. 4.5 tend to indicate that the relationship between SSD and R_eLSPK may actually be non-linear, suggesting that there exists a critical threshold in the rate that SSD increases with R_eLSPK. More data are required for verification, so this idea of a critical threshold will not be discussed further in this study.

4.3.6 P Factor

The main soil conservation techniques used in the Dabie Mountains are contour tillage, contour farming, and level terraces. The support-practice factor (P) is the soil loss ratio when a specific support practice is used, relative to the corresponding

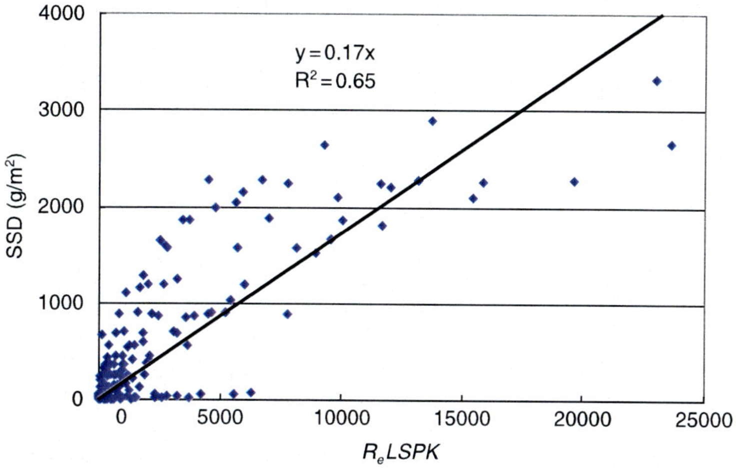

Fig. 4.5 Relationship between $R_e LSPK$ and *SSD* on micro-plots

soil loss that occurs under up-and-down-slope tillage (Renard et al., 1997). In this study, field observations of soil loss with up-and-down-slope tillage at the USLE-plot scale were carried out 200 m away from the sub-catchment of cultivated land from 2000 to 2002, where the crops and human culture activities were almost the same as those in the USLE plot, with level terraces within the sub-catchment of cultivated land. With the field observation data above, the *P* values were calculated. Here, the value of *P* is 0.11.

4.3.7 P_s and C_s Factors

The effect of single operations (which we have also called discrete-event practices) on soil loss in cultivated land is denoted as P_s ($\geqq 1$); C_s is the effect of seasonal coverage on soil loss (0.1–1). We would assume that, using the USLE to predict soil loss from single events without considering the effect of P_s and C_s factors, the predicted values will not agree well with the observed values. It has been observed that the soil loss ratios differ among crops in different growth stages (Gabriels et al., 2003; Fu et al., 2005). In this study, the ratio of observed seasonal soil loss to corresponding predicted seasonal soil loss, as calculated using Eq. (4.9), is defined as $P_s C_s$:

$$P_s C_s = SSD_{ob} / R_e LSKPC_l \tag{4.9}$$

where SSD_{ob} is the total observed *SSD* (g m^{-2}) in the USLE plot in a certain season in the three years from 2000 to 2002; here P_s, C_s, R_e, *L*, *S*, *P*, *K*, and C_l are all the same as those values in Eq. (4.3).

Table 4.1 C_S and P_S factors

Month	Jan., Feb., Nov., Dec.	Mar.	Apr. and May	1st–15th Jun.	15th–30th Jun.	Jul. and Aug.	Sep. and Oct.
$SSD_{ob}(gm^{-2})$	0.00	484.59	1528.16	551.4	4555.11	12911.6	1054.20
Predicted *SSD* with R_eLSKPC_l	3592.82	1475.71	5141.01	1018.99	1983.83	7639.71	1646.76
P_SC_S	0.00	0.33	0.30	0.54	2.30	1.69	0.64
C_S	–	0.33	0.30	0.54	1.00	0.7	0.64
P_S	–	1.00	1.00	1.00	2.30	2.30	1.00

To determine values of P_S and C_S, the field observation data of *SSD* for three years, 2000–2002, from the USLE plot in cultivated land, were used. The seasonal distributions of R_e, *SSD*, and P_SC_S during the three years are listed in Table 4.1.

In Table 4.1, R_e is the effective rain erosivity factor in particular seasons over 3 years from 2000 to 2002; P_SC_S is the ratio of observed seasonal soil loss to corresponding predicted seasonal soil loss; SSD_{ob} is the total observed *SSD* in the USLE plot in particular seasons over the 3 years from 2000 to 2002; P_S is the effect of daily operations on soil loss in cultivated land; and C_S is the effect of seasonal coverage on soil loss.

In April and May, when there is high coverage of wheat and little human activity being carried out in the cultivated land, the minimum P_S value was inferred to be equal to 1. This period is also equivalent to that in September and October, when the corn has matured. In mid-June, when the wheat has been cut and the land is bare, the maximum value of C_S is thought to equal 1. The calculated value of P_S in the period from 15th to 30th of June was 2.3, which was also applied in July and August when grass clearing and land tilling of soil were carried out. During the rainy season from June to July, and in storms in August, there are usually high levels of rain erosivity in the Dabie Mountains, because in this period the bare land during planting of seed by farmers facilitates soil loss. Because it has been reported that mulch with straw or other crop residue can decrease soil loss in cultivated land by 30–98% (Liu et al., 1990; Lu et al., 1988; Wang et al., 1998), the high values of P_SC_S during the summer indicated that mulch on the ground should be utilized to control soil losses. Because of the cold weather in the Dabie Mountains from winter to early spring (November, December, January, and February), field observation data at the USLE-plot scale showed no soil erosion in the cultivated land. Consequently, P_SC_S was thought to be 0 in that period.

4.4 Results and Discussion

Similar to the USLE and RUSLE models, the ER-USLE model only estimates soil erosion from rill and interrill erosion processes. Simanton et al. (1980) applied the USLE to four watersheds and found USLE-estimated soil losses matched reasonably well for two watersheds having no gullies or significant alluvial channels, while

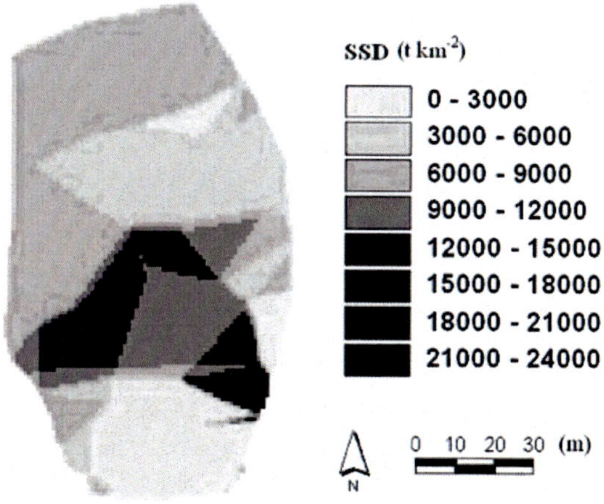

Fig. 4.6 Prediction of annual *SSD* with ER-USLE in year 2000

USLE results did not match well for two watersheds with significant gullies and channels. These results indicate that USLE should not be applied to large watersheds, experiencing significant gully and channel erosion. Erskine et al. (2002) compared RUSLE-estimated soil loss with the measured sediment yield for 12 sub-watersheds in Australia. He did not consider the sediment delivery ratio (the ratio of sediment yield of a drainage basin to the total amount of sediment moved by sheet erosion and channel erosion) in the estimation of soil erosion using RUSL, because the average area for the 12 sub-watersheds was around 5 ha and was 3 ha for the 10 smallest sub-watersheds, where the SDR (sediment delivery ratio) for these small watersheds was high, meaning that most of the soil eroded moved to the downstream areas without significant deposition. In this current study, the size of the sub-catchment of cultivated land is only 0.74 ha, smaller than those reported above. Furthermore, neither gully erosion nor sedimentation was obvious after storm runoff events within the field observation period. Therefore, SDR was not considered in this research.

With the map layer of *LS* and the other factors introduced above, values of annual *SSD* in the year 2000, which were calculated based on single rainfall–runoff events, were inserted for the factors of Eqs. (4.3) and (4.4). The predicted spatial distribution of soil loss in the year 2000 is shown in Fig. 4.6. The values of annual *SSD* predicted by the ER-USLE ranged from 0 to 24,000 (t km^{-2}).

The values of soil loss for single events predicted by both the ER-USLE and the USLE models were calculated and compared with the observed values at the outlet of the sub-catchment of cultivated land and shown in Fig. 4.7. Figure 4.7 shows that the observed values of *SSD* demonstrated a closer linear relationship with the ones predicted by the ER-USLE than those predicted by the USLE, because the USLE over-predicted values of soil loss for some small rainfall–runoff events and for rainfall events when plant coverage was high, while it under-predicted soil loss for events when the cropland had many human activities on it and/or there was low plant coverage.

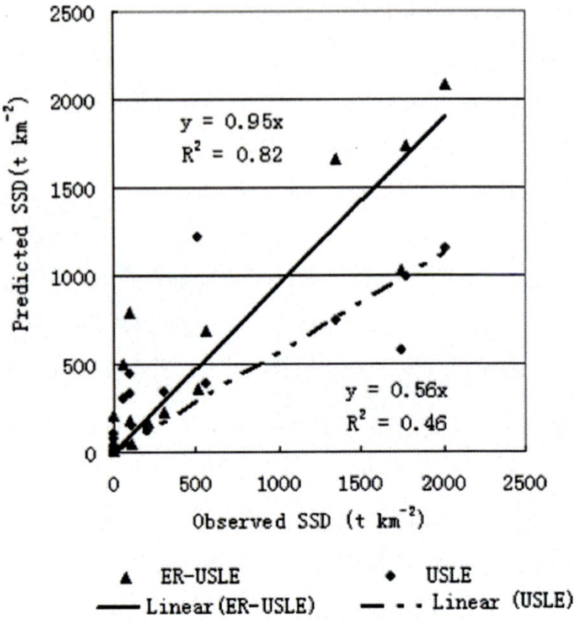

Fig. 4.7 Comparison between observed *SSD* and predicted *SSD* by both ER-USLE and USLE in 2000

In discussing the RUSLE, Renard et al. (1997) proposed that application of USLE caused larger errors because of the seasonal variation of the *K* values. In the current research, we attributed the error to lack of consideration of seasonal variation in the new factors of P_S and C_S, which are closely connected to the seasonal variation of soil loss. Because no comparable field observations were carried out to determine the P_S and C_S factors, the ER-USLE model tried to use the calibration method to calculate the P_S and C_S factors. Values of the calibration result at the USLE-plot scale showed that there were great differences among seasons. The values of P_SC_S were the smallest when the cultivated land was covered with crops and when there were few human activities in March, April, and May, and doubled in September and October. The values of P_SC_S from the middle June through July and August, when the crop was clear-cut and the land was cultivated, were eight times of those in March, April, and May. Because the ER-USLE model was developed from the USLE model, it can be concluded that the ER-USLE model provides a good method for the expansion of USLE and RUSLE models for soil loss prediction based on single events by calculating the different values of P_SC_S for different seasons. These measures are believed to be able to improve the accuracy of USLE model for annual soil loss prediction.

Considering the variation of seasonal coverage of crop and single operations by humans, C_S and P_S are incorporated into the ER-USLE model. Therefore, the ER-USLE model was developed to predict annual soil loss based on single events with an effective R_e factor in the cultivated land. With these measures, it is believed that with the ER-USLE model, annual soil loss in cultivated land can be predicted with high accuracy. Unlike USLE and RUSLE models, which need soil loss data over a long period to predict average annual soil loss, ER-USLE model can be utilized with field observation data for a few years. In addition, the micro-plot method is less expensive and simpler to conduct.

Acknowledgments The work reported in this paper was undertaken within the framework of a research project funded by the Forestry Department of Anhui Province, China. We appreciate the help of Luo Hongyan, Yangyan, Huang Xiaying, Hu Zhidong, Liu Caihong, Wu Zhongneng, Chu Chengdong, and Hu Bingqi for sediment/sample collection and processing. Collaboration of the Forestry Department of Yuexi prefecture in maintaining the monitoring stations at the Shangshe catchment is gratefully acknowledged. We wish to extend our thanks to Prof. Bernard of USDA-Natural Resources Conservation Service (NRCS) for his reading of the manuscript.

References

Bu ZH, Zhao HF, Liu SQ (1993) Preliminary study on algorithm formula of vegetative factor for undisturbed areas in remote sensing monitoring soil loss. Remote Sensing Technol Appl 8:16–22 (in Chinese)

Dickinson A, Collins R (1998) Predicting erosion and sediment yield at the catchment scale. Soil erosion at multiple scales. CAB International, Wallingford, 317–342

Erskine WD, Mahmoudzadeh A, Myers C (2002) Land use effects on sediment yields and soil loss rates in small basins of Triassic sandstone near Sydney, NSW, Australia. Catena 49: 271–287

Fu BJ, Zhao WW, Chen LD, Zhang QJ, Lu YH, Gulinck H, Poesen J (2005) Assessment of soil erosion at large watershed scale using RUSLE and GIS: a case study in the loess plateau of China. Land Degrad Dev 16:73–85

Gabriels D, Ghekiere G, Schiettecatte W, Rottiers I (2003) Assessment of USLE cover-management C-factors for 40 rotation systems on arable farms in the Kemmelbeek watershed, Belgium. Soil Till Res 74:47–53

Griffin ML, Beasley DB, Fletcher JJ, Foster GR (1988) Estimating soil loss on topographically non-uniform field and farm units. J Soil Water Conserv 43:326–331

Jiang ZS, Wang ZQ, Liu Z (1996) Quantitative study on spatial variation of soil erosion in a small watershed in the loess hilly region. J Soil Erosion Soil Conserv 2:1–9 (in Chinese)

Jin ZQ, Shi PJ, Hou FC (1992) Systemic soil erosion model and controlling pattern. Ocean Press, Beijing (in Chinese)

Kinnell PIA (1998) Converting USLE soil erodibilities for use with the Q_REI_{30} index. Soil Till Res 45:349–357

Kinnell PIA (2004) Letter to the editor on "the mathematical integrity of some universal soil loss equation variants." J Soil Sci Soc 68:336–337

Liu YB, Tang KL, Cha X (1990) Experiment on the soil and water loss of slope land with different ground cover. J Soil Water Conserv 4:25–29

Lu ZF, Su M, Li GX (1988) Study on grass and bush and tillage measures to conserve soil and water in loess region. J Soil Water Conserv 2:37–48 (in Chinese)

Ma ZZ (1989) Methods of estimating factors in universal soil loss equation with satellite image. Soil Water Conserv (Chin) 6:24–27 (in Chinese)

Minami N, Yamada T, Nakano M, Tomisaka M, Tokunaga T, Yamashiro O (2002) The characteristics of red sediment discharge at different stages of cultivation of pineapple fields. J Jap Soc Eros Cont (Eng) 54:30–38 (in Japanese)

Moore ID, Burch FJ (1986) Physical basis of the length-slope factor in the Universal soil loss equation. Soil Sci Soc Am J 50:1294–1298

Neering MA (1998) Why soil erosion models over-predict small soil losses and under-predict large soil losses. Catena 32:15–22

Renard KG, Foster GR, Weesies GA, McCool DK, Yoder DC (1997) Predicting soil erosion by water: a guide to conservation planning with the Revised Universal Soil Loss Equation (RUSLE). In: Agriculture Handbook, vol 703. US Department of Agriculture, Washington, DC

Simanton JR, Osborn HB, Renard KG (1980) Application of the USLE to southwestern rangelands. Hydro Water Res Ariz SW 10:213–220

Wang CH, Xiao J, Wang ZG (1998) A preliminary study on effects of straw mulches on surfaces flow and soil erosion on slope farmland. J Shanxi Agric Univ 18:149–152

Williams JR (1972) Sediment-yield prediction with universal equation using runoff energy factor. In: Present and prospective technology for prediction sediment yield and sources: proceedings of the sediment yield workshop. USDA Sedimentation Lab, Oxford, MS, ARS-S-40, pp 244–252

Williams JR, Berndt HD (1972) Sediment yield computed with Universal equation. J Hydrol Div ASCE 98:2087–2098

Williams JR, Berndt HD (1977) Sediment yield prediction based on watershed hydrology. Trans ASAE vol 20(6):1100–1104

Williams JR, Renard KG, Dyke PT (1984) A new method for assessing erosion's effect on soil productivity. J Soil Water Conserv 38:381–383

Wilson JP (1986) Estimating the topographic factor in the Universal Soil Loss Equation for watershed. J Soil Water Conserv 41:179–184

Wischmeier WH, Smith DD (1965) Predicting rainfall erosion losses from cropland east of the Rocky Mountains: guide for selection of practices for soil and water conservation planning. USDA Agriculture Handbook, US Government Printing Office, Washington, DC

Wischmeier WH, Smith DD (1978) Predicting rainfall erosion losses. A guide to conservation planning. In: Agricultural handbook, vol 537. US Department of Agriculture, Washington, DC

Yang KB, Guo PC (1994) Study of erosivity index of rainfall in Northern Shanxi hilly and gully area. Bull Soil Water Conserv 14:31–35 (in Chinese)

Zhang JC, Hu HB (1996) Water and soil loss in China. Chinese Forestry Press, Peking (in Chinese)

Zhang JC, Ruan HH, Hu HB et al (1999) The present situation and control measures of soil and water loss in Changjiang River Valley. J Nanjing Forestry Univ 23(2):18–22

Zhang Y, Yuan JP, Liu BY (2002) Advance in researches on vegetation cover and management factor in the soil erosion prediction model. Chin J Appl Ecol 13:1033–1036

Zhang JC, Zhaung JY, Nakamura H, Cheng P (2004) Suspended sediment discharges from various land uses in the Shangshe catchment in the Mountains, China. Ninth international symposium on river sedimentation, Yichang, China. Qinhua Univeristy Press, Beijing, pp 2426–2431

Chapter 5
Development and Test of GIS-Based FUSLE Model in Sub-catchments of Chinese Fir Forest and Pine Forest in the Dabie Mountains, China

Abstract Land use plays an important role in soil loss and other environmental problems. Correct prediction of soil loss from different types of land use is very important to land use policy making in the Dabie Mountains, China. Field observations of water and soil loss were carried out in the Shangshe catchment on land in four types of use from 1999 to 2007. This chapter reports the study of soil loss in the sub-catchment of Chinese fir forest and the sub-catchment of pine forest. Field observations of water and soil loss were carried out at the micro-plot scale, the USLE (universal soil loss equation)-plot scale, and the sub-catchment scale in the sub-catchment of Chinese fir forest, as well as of pine forest. Analysis of these field observation data shows that litter in forest has an important hydrological function. In the Chinese fir forest and pine forest, the micro-plots without litter and grass produced more than 71 times the soil loss of micro-plots with litter and grass at the same gradient in 2000. By integrating a linear regression method with GIS and USLE, a so-called FUSLE (USLE in forest with a focus on litter) model was developed to predict soil loss in forest. The rain erosivity factor is turned into a modified rain erosivity factor when litter is added as a new factor. These measures are believed more practical for soil loss prediction in forest areas because the litter factor is a key factor in soil loss on forestland. Furthermore, the meter-scale plot method is able to get enough field observation data in a few years for soil loss prediction and it is less expensive.

5.1 Introduction

Soil erosion has increased throughout the 20th century in the watershed of the Yangtze River, China (Zhang et al., 1999). It has caused sedimentation at dams and in riverbeds, which in turn has intensified flooding. Much effort has been made to understand the mechanisms of soil erosion and to predict soil loss. The universal soil loss equation (USLE) (Wischmeier and Smith, 1965) is an empirical model for predicting the average annual soil loss caused by rainfall. With this model, soil loss is evaluated by extrapolating from plot and sub-catchments to catchments. Attempts have been made to use it on forestland (Wischmeier and Smith, 1978; Lufafa et al., 2003) and pasture (Bacchi et al., 2003). However, in dense forest ecosystems, canopy coverage is very high. For the purpose of considering single rainfall events in an evergreen forest, canopy coverage does not vary much among

J.C. Zhang et al., *Theory and Practice of Soil Loss Control in Eastern China*,
DOI 10.1007/978-1-4419-9679-4_5,

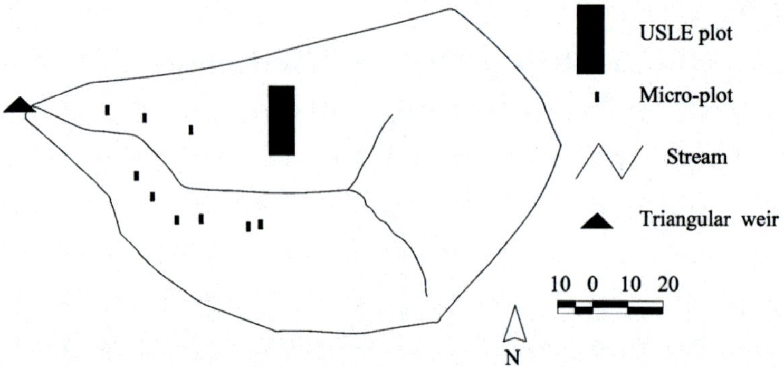

Fig. 5.1 Sub-catchment of Chinese fir forest showing the locations of the USLE plot, the micro-plots, and the triangular weir

seasons; so seasonality of effects may not be strong. Moreover, usually the forest floor is covered with litter and grass, which have good buffering effects on overland flows, because litter has a great detention and retention capacity (Balci, 1963; Wu and Wang, 1993). The major obstacle of application of the USLE approach may be the spatial heterogeneity in the forest catchment, and additional methods need to be tried for soil loss prediction in forests.

Accurate estimation of soil erosion due to water is very important in several environmental contexts, such as the assessment of potential soil loss from forests and the evaluation of the loss of water storage capacity in reservoirs due to sediment deposition. To predict the response of watersheds and sub-systems, quantification of their hydrological and erosion behavior is needed. In the Shangshe catchment, there are five main kinds of land use: paddy field, cultivated land, pine (*Pinus massoniana*) forest, Chinese fir (*Cunninghamia lanceolata*) forest, and tea (*Camellia sinensis*) garden. In the Shangshe catchment, four sub-catchments of different land types were selected as experimental sites (Fig. 5.1). Four 45° sharp-crested, V-notch weirs were set up in the catchments of pine forest (0.86 ha), Chinese fir forest (0.89 ha), cultivated land (0.74 ha), and tea garden (0.59 ha), respectively. This study was carried out in the sub-catchment of Chinese fir forest.

Soil erosion monitoring can be conducted on-site (at plot level) and off-site (at sub-catchment and catchment levels). The advantages and disadvantages of these two monitoring approaches have been discussed in the literature. Many studies of soil erosion have been conducted at sub-catchment or catchment levels: this approach can better describe the rainfall–runoff response at the catchment scale and response to certain management practices. Upslope or on-site monitoring, on the other hand, is relatively simple to conduct and inexpensive. This type of monitoring is best suited to portraying soil erosion processes and soil disturbances on-site (Corner et al., 1996). Comparison of simulated erosion patterns with observed erosion patterns is necessary to confirm a model's robustness, not only for the soil erosion monitoring off-site but also for on-site soil erosion. Analysis of the soil erosion by both off-site and on-site soil erosion monitoring will show their mutual relationship. Because of these considerations, both the on-site and off-site soil erosion monitoring methods were used.

5.2 Study Area

This study was conducted in the Shangshe catchment (34°32′20″N,116°50′12″E) of the Dabie Mountains in Yuexi prefecture of Anhui Province, China. Details of the Shangshe catchment were reported by Zhuang et al. (2004). Within the Shangshe catchment, a sub-catchment of Chinese fir forest of 8900 m^2 was selected for field observations (Fig. 5.1). The average height of trees was 6.1 m and the $\overline{D}_{1.3}$ (average diameter of trees at the 1.3 m height) was 8.1 cm. The tree density was 1660 trees ha^{-1}. Under the forest, there were grasses, shrubs, and litter.

This study investigated soil loss at the micro-plot level, the USLE-plot level, and the sub-catchment level, with the following objectives: (1) to develop an empirical soil loss model suitable for predicting soil loss from single rainfall–runoff events in the forest and (2) to investigate the extent to which on-site monitoring of soil losses agrees with the results of off-site monitoring of sediment yields in the forest.

5.3 Materials and Methods

5.3.1 Field Observation at the Sub-catchment Scale

A 45° V-notch, sharp-crested weir, equipped with a float-type water level recorder, was built at the outlet of the sub-catchment of the Chinese fir forest (Fig. 5.1). At 8 a.m. each day, the recorder paper of a water gauge was changed manually; the water level at the weir was measured in order to correct possible errors of the water level in the recorder paper. Details of water runoff and suspended sediment concentration observations, as well as runoff and suspended sediment discharge calculations at the sub-catchment scale, are described in Zhang et al. (2004). The sediment at the weir was measured every 3 months and then an amount attributed to each event according to its ratio of suspended sediment discharge in each season. The sediment discharge (SD, t) was transformed into specific sediment discharge (SSD, t km^{-2}). Field observation data of SSD for single events in the year 2000 were used in this study.

5.3.2 Precipitation Observation

A manual rain gauge and an autographic rain gauge were set up to survey precipitation and rainfall intensities 400 m away from the weir at the monitored site of Chinese fir forest. The recorder paper of the autographic rain gauge was changed daily at 8 a.m. For every rainfall event, precipitation was read from the recorder paper for the automatic rain gauge and compared with data from the manual rain gauge. Field observation data of precipitation in 1999–2002 were used in this study.

5.3.3 Field Observations at the USLE-Plot and Micro-plot Scales

The method of micro-plots at a meter scale was applied in the sub-catchment of Chinese fir forest in year 2000. Nine micro-plots of 2 m × 1 m on slopes of 3°, 6°, 8°, 11°, 15°, 25°, 28°, 33°, and 35° were set up, whereas a USLE plot of 22.1 m × 5 m was set up on a slope of 27.9° in the sub-catchment of Chinese fir forest (Photo 5.1). The micro-plots of 3°, 6°, 8°, 11°, and 15° were covered with natural litter and grass, while the litter and grass on the other four micro-plots were removed. For each micro-plot, one container was used to collect water and sediment discharge, whereas for the USLE plot, the volume of surface runoff was calculated by measuring the height of the water in the first and second collecting tanks. A 200–500 ml sample was taken to the laboratory, where the sediment was filtered, dried in an oven at 110°C, and weighed. For each rainfall–runoff event, the runoff volume and the soil sediment loss from the micro-plots and the USLE plot were calculated. The sediment discharge (SD, t) was transformed into SSD (t km^{-2}). Field observation data of the USLE plot in 2000, 2001, and 2002 were used in this study.

Photo 5.1 A USLE plot in a Chinese fir forest

5.3.4 FUSLE Model for Soil Loss Prediction

The USLE has been applied widely at a watershed scale on the basis of an aggregated approach (Williams and Berndt, 1972, 1977; Wilson, 1986; Griffin et al., 1988; Dickinsonou and Collins, 1998). GIS techniques, which reduce the time of analyses, have facilitated the application of USLE with a spatially distributed approach (Kinnell, 2001). However, forested land is a completely different ecosystem from that of cultivated land. Canopy cover, sapling density, litter depth, and woody debris appear to be important ecological factors that determine the magnitude of soil loss (Herlina et al., 2003). Trees provide and maintain a litter layer, which protects the soil against the impact of raindrops (Binkley and Brown, 1993). Wang and Xie (1998) reported that soil cover removal often increases erosion by 10–100 times, while tree canopy removal, when the soil cover is not disturbed, increases soil erosion rate by less than 50% (Wang and Xie, 1998; Herlina et al., 2003). The protective values of stands consist in both the ability of the canopy to decrease the power of raindrops and their ability to provide materials for soil cover on the forest floor. As a result, the energy of raindrops, depending on their drop size and velocity, is reduced to almost zero when they reach the soil (Binkley and Brown, 1993). Because of the powerful function of litter on soil loss control in forest, the litter is added as a factor into the USLE model, thereby creating what is called a FUSLE model (USLE in forest with a focus on litter), as in Eq. (5.1):

$$Y = CR_{\mathrm{e}}LSPKl_t \tag{5.1}$$

where

Y is the SSD (t km^{-2} or g m^{-2});
L is the slope length factor;
S is the slope factor;
P is the support-practice factor;
C is the vegetation cover and management factor;
l_t is the litter factor;
R_{e} is the effective rain erosivity factor (J-mm m-2 h-1); and
K is the soil erodibility factor (g h J-1 mm-1).

5.4 Factors in FUSLE Model

5.4.1 R_e Factor

Yang and Guo (1994), using observed data from normal runoff plots in cultivated land, analyzed the relationship between the R values calculated at different intervals (5 min to 1 h) and the amount of soil loss for every event. They reported that the soil loss at the USLE-plot scale was closely correlated with R values calculated at 10-min intervals in northern China. This method is applied in the current study. The equation for R is (Wischmeier and Smith, 1965)

$$R = \sum_{i}^{n} E_i I_{30} \tag{5.2}$$

where

R is the rainfall–runoff erosivity factor by precipitation (J mm m^{-2} h^{-1});
E_i is the rain energy in one storm in i 10-min intervals; and

I_{30} is the maximum continuous rain intensity in a 30-min period in a storm (mm h^{-1}).

In China, the R factor is reduced by a factor of 0.01 in application.

The Committee of the Yellow River of China has also computed a regression equation that incorporates rain energy and intensity (Zhang and Hu, 1996):

$$E = 210.3 + 89 \log_{10} I \tag{5.3}$$

where

E is the rain energy in one period (J m^{-2} cm^{-1}) and
I is the rain intensity (cm h^{-1}).

This regression equation was applied to calculate E in this study.

In general, erosion models have difficulty predicting small-scale events because of the large natural variation of factors related to soil loss (Neering, 1998). Effective rainfall erosivity was proposed as a method to smooth the errors in rain loss on soil loss in the USLE application in China. It is calculated from Eq. (5.4):

$$R_e = R - R_l \tag{5.4}$$

where

R_e is the modified R factor (J mm m^{-2} h^{-1});
R is the rain erosivity factor (J mm m^{-2} h^{-1}); and
R_l represents the loss of rainfall erosivity, or reduction of the R factor, up until the occurrence of rainfall–runoff (J mm m^{-2} h^{-1}).

In 1999 and 2000 in the Shangshe catchment, 28 and 18 rainfall–runoff events, respectively, occurred in the sub-catchments of Chinese fir. Because runoff observation was also carried out in the sub-catchment of Chinese fir, by comparing the hydrograph of runoff with the hyetograph of precipitation, the loss of rainfall could be determined. Equation (5.5) was applied to calculate the R_e in this study:

$$R_e = (E - E_l) I_{30} \tag{5.5}$$

where

R_e is the effective R factor (J mm m^{-2} h^{-1});
E is the rain energy of one rainfall–runoff event (J m^{-2}); and
E_l is the rain loss energy that takes place until the occurrence of rainfall runoff in the same rainfall–runoff event (J m^{-2}).

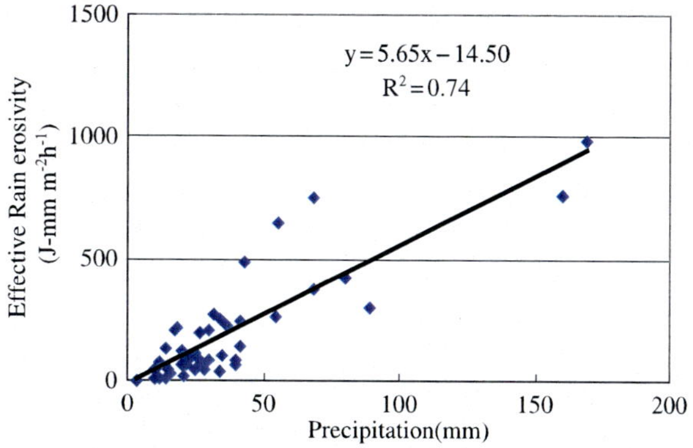

Fig. 5.2 Relationship between rainfall and R_e in a Chinese fir forest

For the 46 rainfall-runoff events in 1999 and 2000, the values of R_e were calculated. R_e was linearly related to rainfall in the Chinese fir forest (Fig. 5.2). A linear regression equation of R_e with the precipitation was determined:

$$R_e = 5.65x - 14.5 \tag{5.6}$$

where x represents the rainfall (mm) for single events.

5.4.2 L and S Factors

5.4.2.1 Effect of Slope on Soil Loss at the Micro-plot Scale

Analyses of field observation data of 15 rainfall–runoff events during year 2000 at the micro-plot scale showed that WD (water discharge, mm) and SSD increased with the slope, despite the fact that the relationship between slopes and SSD at the micro-plot scale varied among events. The 16 August 2000 event is shown as an example (Fig. 5.3). The values of SSD in the micro-plots with slopes of 25°, 28°, 33°, and 35° were extremely low. This was due to the effect (or function) of litter and grass on the forest floor, which will be discussed below.

5.4.2.2 *LS* Factors at the Sub-catchment Scale

The slope was derived from the digital elevation model (DEM) of the sub-catchment of Chinese fir forest at the 1 m × 1 m grid size (Fig. 5.4). In both of the USLE and RUSLE models, slope length is defined as the horizontal distance from the origin of the overland flow to the point where either the slope gradient decreases to a point at which deposition begins or runoff becomes concentrated in a defined channel (Wischmeier and Smith, 1978; Renard et al., 1997). In the present study, the *L* factor was determined using the following method. First, the flow path was determined under the hydro model using ArcView GIS for the Chinese fir forest sub-catchment. Then, the sub-catchment of Chinese fir forest was divided into three

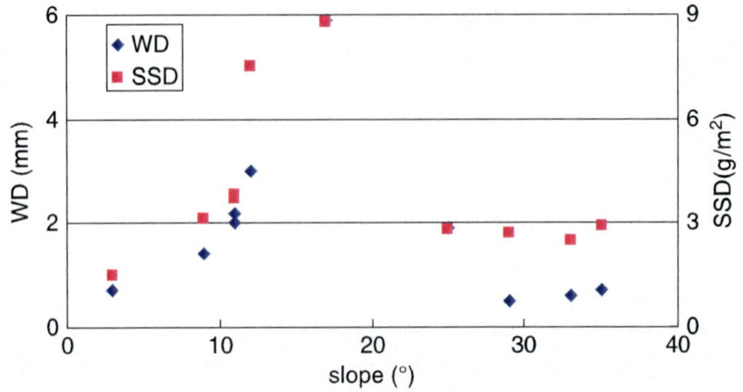

Fig. 5.3 Relationship among WD, SSD, and slope at the micro-plot scale on 16 August 2000

Fig. 5.4 DEM of the sub-catchment of Chinese fir forest

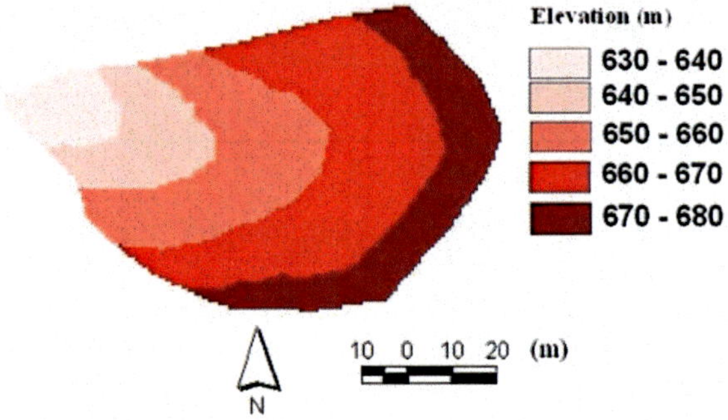

parts along a main stream and two sub-streams (Fig. 5.1). The slope length in each of the three parts is equal to the ratio of its areas to the length of its stream.

The *LS* factor was calculated according to Eq. (5.7) (Moore and Burch, 1986) and is shown in Fig. 5.5:

$$LS = (\lambda/22.1)^{0.6}\left(\frac{\sin\theta}{0.0896}\right)^{1.3} \tag{5.7}$$

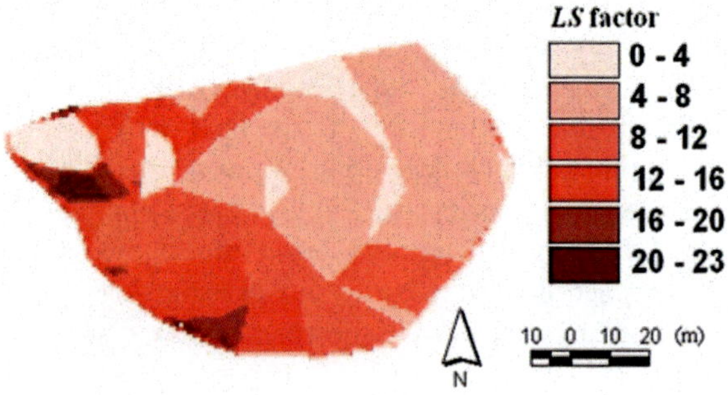

Fig. 5.5 *LS* factor in the sub-catchment of Chinese fir forest

where

LS is a unitless terrain factor;
λ is the length of the slope in m; and
θ is the slope in degrees.

5.4.3 K Factor

K is calculated using Eq. (4.8).

5.4.4 C Factor

The vegetation cover and management factor (C) in the USLE has been studied extensively and used widely in the USA. For a given type of land use, it is assumed that values of SSD at the micro-plot scale show a linear regression relationship with R_eLSPK. The constant parameter of the linear regression model between SSD and R_eLSPK is defined as C. The P factor in the forest is assumed to be 1.0. This method is used to determine the C factor in this study.

Field observation data of 15 rainfall-runoff events at the micro-plot scale were used for parameter estimates of the linear regression. The SSD at the micro-plot scale without litter showed a linear relationship with R_eLSPK(Fig. 5.6). In Chinese fir forest, C is equal to 0.0827.

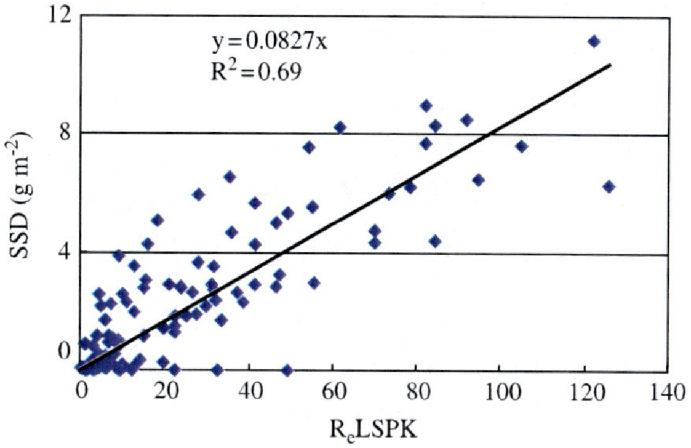

Fig. 5.6 Linear regression of SSD with R_eLSPK

5.4.5 Litter Factor

Trees provide and maintain a litter layer, which protects the soil against the impact of raindrops (Binkley and Brown, 1993). Litter decomposition is vital to nutrient cycling and the productivity of forests (Didham, 1998) and is an important component of the global budget (Aerts, 1997). On the other hand, the hydrological function of the leaf litter layer is also powerful. The tree canopy determines the size and erosive power of the raindrops. Saplings, grass, the litter layer, and woody debris protect the soil surface, thus preventing soil detachment, and provide surface roughness that minimizes soil particle movement down the slope (Wang and Xie, 1998; Herlina et al., 2003). The effect of litter on soil erosion control was reported in the 1930s (Lowdermilk, 1930; Corter, 1938). In China, work in measuring the effect of litter began in the 1960s. Especially during the last 20 years, the hydrological effect of litter had been reported in most forest types all over China. Tan and Zhang (1995) reported that the water-holding capacity of litter in the Chinese fir and maple (*Liquidambar formosana*) forest of the lower reaches of the Wu River was about 3 mm. Similarly, Yu et al. (2002) reported that the effective maximum water-holding capacity of litter and moss in Emei fir (*Abies fabri*) forests can reach 3.23 mm. The water regulating function of litter and moss will increase with the Emei fir forest succession. Rao et al. (2005) concluded that the effective maximum water-holding capacity of forest litter in the Simian Mountain in Sichuan Province was between 1.8 and 4.6 mm. It is well accepted that roots stop rill development, while the litter layer limits splash erosion.

Plots with litter produced much lower *SSD* than did the plots without litter (Fig. 5.3). The other rainfall events in 2000 showed similar results. This means that small areas without litter within catchments can be responsible for most of the runoff production, which also has implications for sediment delivery. Soil loss in a forest catchment should be the total sum of soil losses from areas both with litter and without litter. Therefore, at the catchment scale or sub-catchment scale, the litter coverage should also be included in the litter factor to predict soil loss. Equation (5.8) is used to calculate the litter factor:

$$l_t = L_i * \mathrm{Cov}_{Li} + 1 - \mathrm{Cov}_{Li} \tag{5.8}$$

where

l_t is the litter factor;

L_i is the ratio of annual *SD* at a micro-plot with litter to *SD* at the micro-plot of the same gradient without litter (here it is 0.014); and

Cov_{Li} is the coverage of litter in the sub-catchment.

$L_i * \mathrm{Cov}_{Li}$ means the coefficient of SSD from the area in a sub-catchment with litter, and $1 - \mathrm{Cov}_{Li}$ means the coefficient of SSD from the other area in the sub-catchment without litter.

Field observations of the spatial distribution of litter coverage were carried out in March 2000 in the sub-catchment of Chinese fir forest. Using Eq. (5.8), the spatial distribution of litter factor was calculated and is shown in Fig. 5.7. In Fig. 5.7, the areas where the values of the litter factor are between 0.62 and 0.76 are the roads and other places without litter, grass, or shrubs on the ground.

5.5 Results

With the map layer of *LS* (Fig. 5.5) and other factors introduced above, the temporal and spatial soil losses for 15 events in 2000 were calculated using Eq. (5.1). Predicted values of SSD at the sub-catchment scale with the FUSLE model for

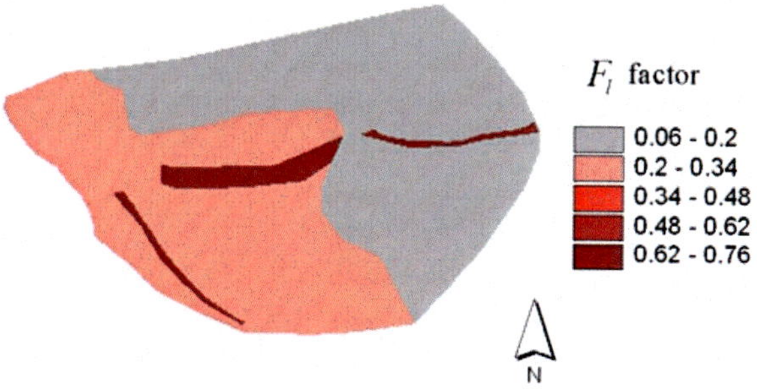

Fig. 5.7 Spatial distribution of litter factor

single events were compared with the observed values of *SSD* at the monitoring site of the sub-catchment outlet (Fig. 5.8). Their R^2 value is 0.73. The spatial distribution of predicted annual soil loss in 2000 was transformed into a graph (Fig. 5.9). The values varied between 0 and 690 (t km^{-2}) in the sub-catchment of the Chinese fir forest. The results are similar to the results in the sub-catchment of pine forest and tea garden (two studies in the Shangshe catchment, unpublished). Predicted values of annual *SSD* at the USLE-plot scale with the FUSLE model in the 3 years of 2000, 2001, and 2002 agreed well with the observed values of annual *SSD* (Fig. 5.10). Annual values of *SSD* of 2000, 2001, and 2002 at the USLE-plot scale varied from 4 to 6.5 t km^2. They were very low because of the high coverage of litter and shrubs in the USLE plot.

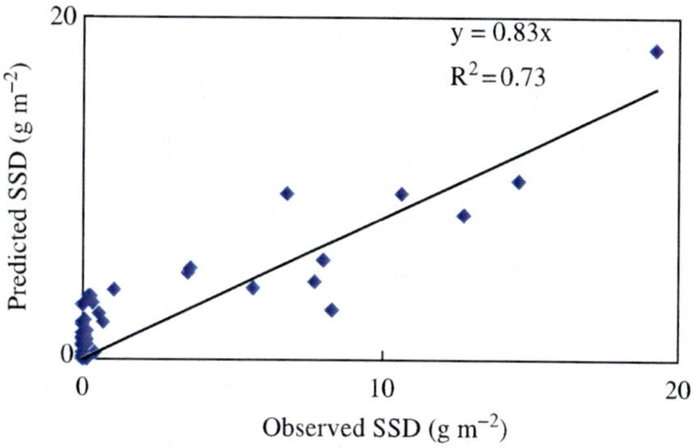

Fig. 5.8 Comparison between observed values of SSD for single events and predicted values at sub-catchment scale in 1999 and 2000

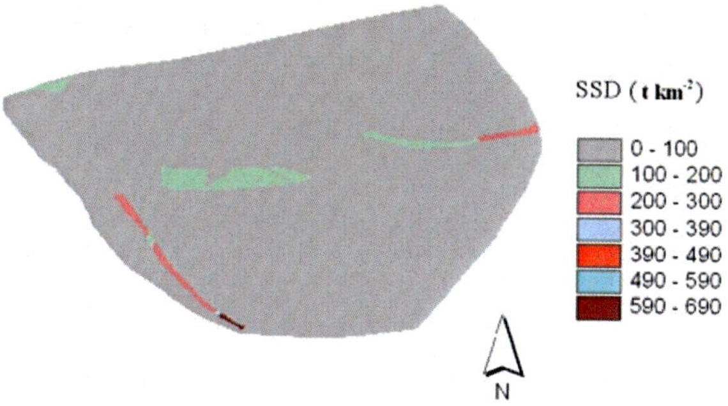

Fig. 5.9 Spatial distribution of predicted annual soil loss in 2000

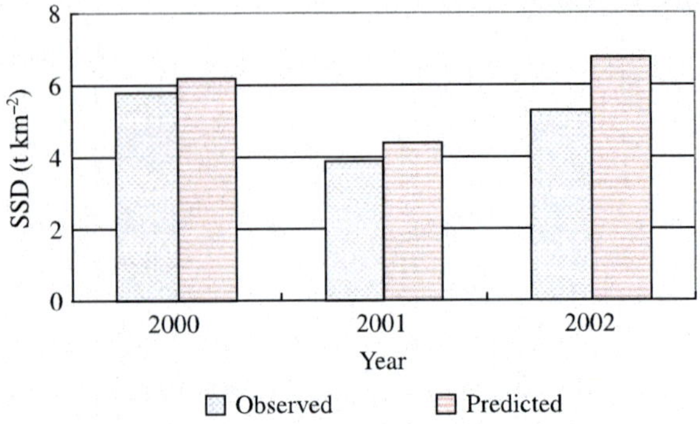

Fig. 5.10 Comparison between observed values of annual *SSD* at the USLE-plot scale with predicted ones in 2000, 2001, and 2002

5.6 Application and Test of FUSLE in a Sub-catchment of Pine Forest

5.6.1 Materials and Field Observations of Runoff and Soil Loss in the Sub-catchment of Pine Forest

Similar to the monitoring sites in the sub-catchment of Chinese fir forest, a sub-catchment of pine forest of 8600 m^2 was selected in the Shangshe catchment (Fig. 5.11), where the average height of trees was 6.2 m and the average diameter of trees at 1.3 m height was 10.1 cm. The density of trees was 1250 ha^{-1}. Both grass and litter were on the ground layer of the forest. A 45° V-notch, sharp-crested weir, equipped with a float-type water level recorder, was built at the outlet of the sub-catchment of pine forest. The method of micro-plots at a meter scale was applied near the sub-catchment of pine forest in the year 2000. The pine trees and litter and grass of the micro-plots were similar to those in the sub-catchment of pine forest. Eight micro-plots of 2 m × 1 m on slopes of 12°, 12°, 17.8°,17.8°, 22°, 23°, 23°, and 27.5° were set up, while a USLE plot of 22.1 m × 5 m was set up on a slope of

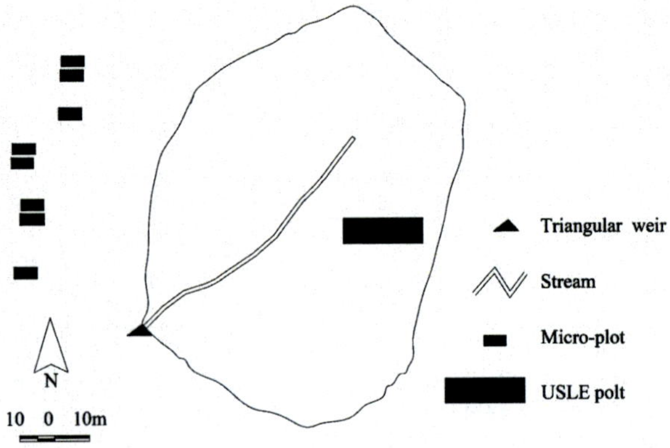

Fig. 5.11 USLE plot and the triangular weir in the sub-catchment of pine forest and the micro-plots outside of it

27.9° in the sub-catchment of pine forest (Fig. 5.11). The micro-plots of 12°, 17.8°, 22°, and 23° were covered with natural litter and grass, while the litter and grass on the other four micro-plots were removed. Field observations at the micro-plot scale, USLE-plot scale, and sub-catchment scale were made in the same way as those in the sub-catchment of Chinese fir forest.

5.6.2 Application and Test of FUSLE in the Sub-catchment of Pine Forest

5.6.2.1 Factors in the FUSLE Model

(1) R_e factor
The R_e factor calculated above for the sub-catchment of Chinese fir forest was used in this research on pine forest.
(2) *L* and *S* factors

(a) Effect of slope on soil loss at the micro-plot scale
Analyses of field observation data of 15 rainfall–runoff events in the year 2000 at the micro-plot scale showed that *WD* (water discharge, mm) and *SSD* increased with the slope, despite the fact that the relationship between slopes and *SSD* at the micro-plot scale varied among events. The 22 June 2000 event is shown as an example (Fig. 5.12). The values of *SSD* on the micro-plot of plots with slopes of 12°, 17.8°, 22°, and 23°, which were covered with natural litter and grass, were extremely low. This was because of the properties (or ecosystem function) of litter and grass on the forest floor, which will be discussed later.
(b) *LS* factors at the sub-catchment scale
The slope was derived from the digital elevation model (DEM) of the sub-catchment of pine forest at the 1 m×1 m grid scale (Fig. 5.13). In the present study, the *L* factor was determined using the following method.

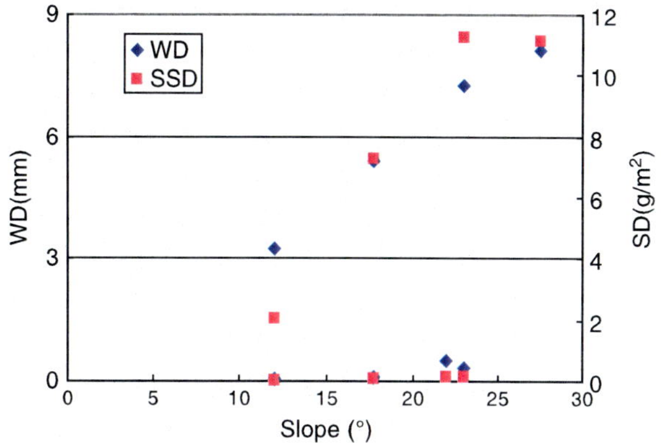

Fig. 5.12 Relationship among *WD, SSD,* and slope at the micro-plot scale on 22 June 2000

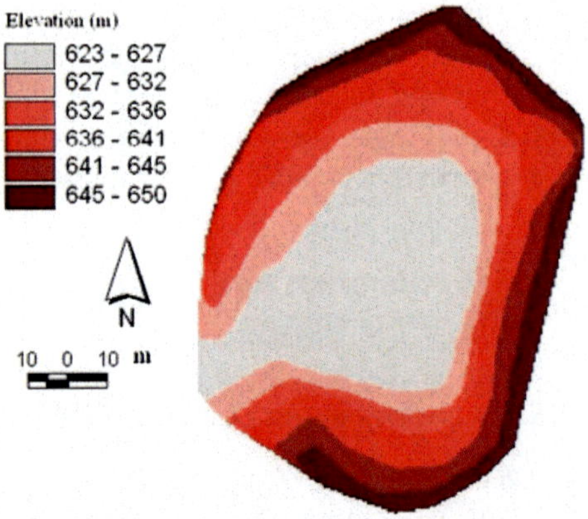

Fig. 5.13 DEM of the sub-catchment of pine forest

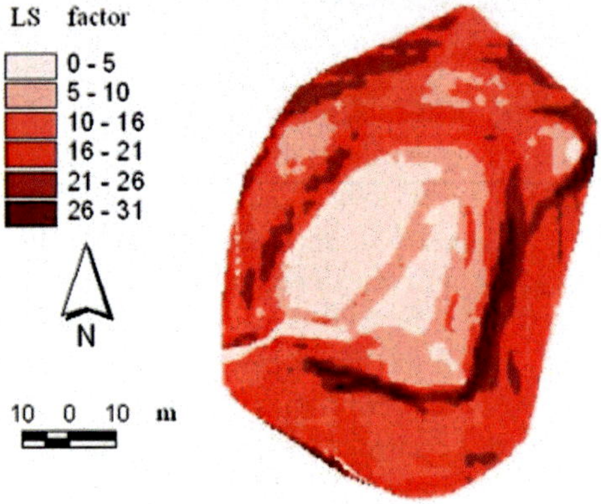

Fig. 5.14 *LS* factor in the sub-catchment of pine forest

> First, the flow path was determined using the hydro model under ArcView GIS for the pine forest sub-catchment. Then, the sub-catchment of pine forest was divided into two segments along the stream (Fig. 5.11). The slope lengths in the two side segments are calculated as the ratios of their areas to the length of stream.

The *LS* factor was calculated according to Eq. (5.7) and is shown in Fig. 5.14.

5.6.2.2 *K* Factor

K is calculated with Eq. (4.8).

5.6.2.3 *C* Factor

The vegetation cover and management factor (*C*) in the USLE has been studied extensively and used widely in the USA. For one given type of land use, it is assumed that values of *SSD* on micro-plots will show a linear regression relationship with R_eLSPK. The constant parameter of the linear regression model between *SSD*

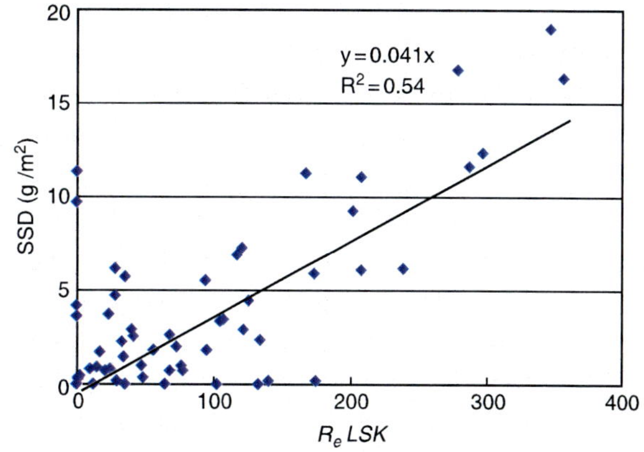

Fig. 5.15 Linear regression of *SSD* with R_eLSPK

and R_eLSPK is defined as the value of *C*. The *P* factor in the forest is assumed to be 1.0. This method is used to determine the *C* factor in this study.

Field observation data of 15 rainfall-runoff events at the micro-plot scale were used for estimates of parameter *C* from linear regression. The *SSD* at the micro-plot scale without litter can be described by a linear relationship with R_eLSPK (Fig. 5.15). In pine forest, *C* equals to 0.041.

5.6.2.4 Litter Factor

Field observations of litter coverage were made in the sub-catchment of pine forest and Eq. (5.8) was used to calculate the litter factors in the years 2000 and 2007, which are shown in Figs. 5.16 and 5.17, respectively. In Fig. 5.16, the values of the litter factor over the area in the year 2000 were between 0.06 and 0.21. In Fig. 5.17, the litter factor over the area in the year 2007 decreased to the range of 0.05–0.11, because the coverage of shrubs and small plants increased on the forest floor.

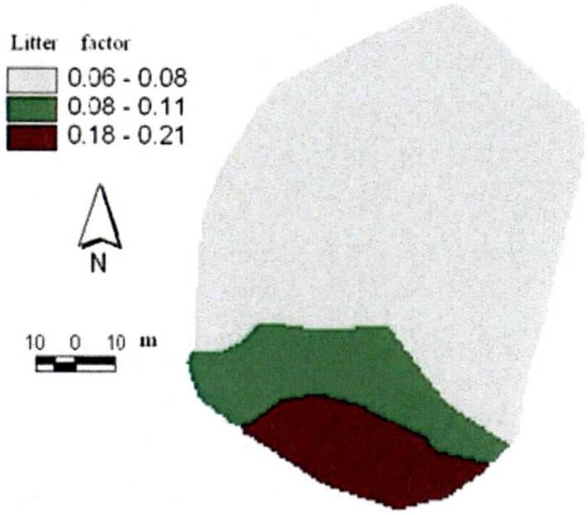

Fig. 5.16 Spatial distribution of predicted annual litter factor in year 2000

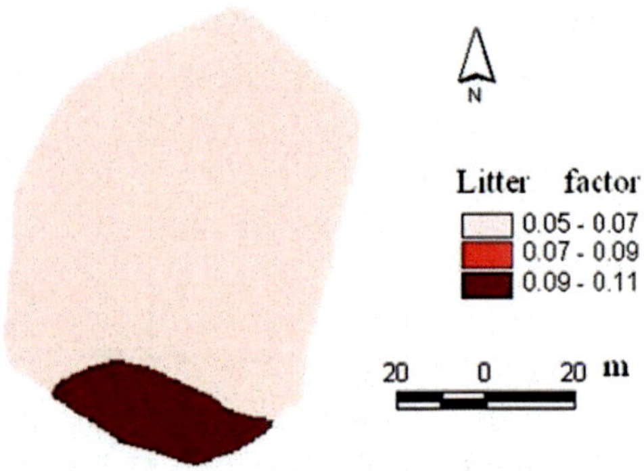

Fig. 5.17 Spatial distribution of the predicted litter factor in year 2007

With the map layer of *LS* (Fig. 5.14) and other factors introduced above, the temporal and spatial soil losses for 15 events in 2000 and several events in 2007 were calculated using Eq. (5.1). Spatial distributions of the predicted annual soil losses in the years of 2000 and 2007 were transformed into graphs (Figs. 5.18 and 5.19). The values varied between 0 and 37 (t km^{-2}) in the sub-catchment of pine forest in the year 2000 and decreased to 0–20 (t km^{-2}) in the year 2007 with almost the same effective rainfall erosivity as that in the year 2000. Predicted values of annual SSD at the sub-catchment scale with the FUSLE model in the 8 years of 2000–2007 were compared with the observed values of SSD at the monitoring site of the sub-catchment outlet of pine forest (Fig. 5.20). Their R^2 value is 0.80. Predicted values of annual SSD at the USLE-plot scale with the FUSLE model in the 8 years of 2000–2007 agreed well with the observed values of annual SSD, which varied from 2.1 to 6.5 t km^{-2}. These values were very low because of the high coverage of litter and shrubs in the USLE-plot.

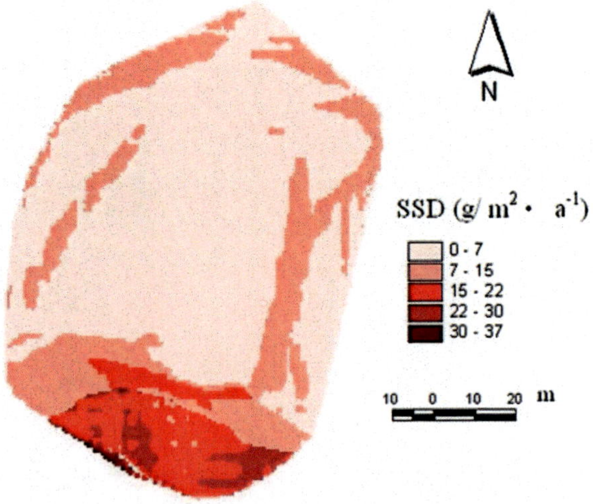

Fig. 5.18 Spatial distribution of predicted annual soil loss in 2000

Lowdermilk WC (1930) Influence of forest litter on runoff protection and erosion. J Forest 28: 474–491

Lufafa A, Tenywa MM, Isabirye M, Majaliwa MJG, Woomer PL (2003) Prediction of soil erosion in a Lake Victoria basin catchment using a GIS-based Universal Soil Loss model. Agric Syst 76:883–894

Moore ID, Burch FJ (1986) Physical basis of the length-slope factor in the Universal soil loss equation. Soil Sci Soc Am J 50:1294–1298

Neering MA (1998) Why soil erosion models over-predict small soil losses and under-predict large soil losses. Catena 32:15–22

Rao LY, ZHU JZ, Bi HX (2005) Hydrological effects of forest litter and soil in the Simian Mountain of Chongqing City. J Beijing Forestry Univ 27:33–38 (in Chinese)

Renard KG, Foster GR, Weesies GA, McCool DK, Yoder DC (1997) Predicting soil erosion by water: a guide to conservation planning with the Revised Universal Soil Loss Equation (RUSLE). In: Agriculture handbook, vol 703. US Department of Agriculture, Washington, DC

Tan LY, Zhang WG (1995) Research on the hydrological properties of forest litter and its different decomposition layer in the lower reach of Wujiang River. J Southwest Forestry Coll 15:32–38 (in Chinese)

Wang LX, Xie MS (1998) Study on the hydrological effect of protection forest in mountain area. Chinese Forestry Press, Peking, China (in Chinese)

Williams JR, Berndt HD (1972) Sediment yield computed with universal equation. Proc Am Soc Civil Eng 98(HY12):2087–2098

Williams JR, Berndt HD (1977) Sediment yield prediction based on watershed hydrology. Trans ASAE 20(6):1100–1104

Wilson JP (1986) Estimating the topographic factor in the Universal Soil Loss Equation for watershed. J Soil Water Conserv 41:179–184

Wischmeier WH, Smith DD (1965) Predicting rainfall erosion losses from cropland east of the Rocky Mountains: guide for selection of practices for soil and water conservation planning. USDA Agriculture Handbook. US Government Printing Office, Washington, DC

Wischmeier WH, Smith DD (1978) Predicting rainfall erosion losses, a guide to conservation planning. In: Agriculture handbook, vol 537. US Department of Agriculture, Washington, DC

Wu CW, Wang LX (1993) The function of litter on soil loss controlling. Chinese Water Soil Loss Controlling 4:55–61 (in Chinese)

Yang KB, Guo PC (1994) Study of erosivity index of rainfall in Northern Shanxi hilly and gully area. Bull Soil Water Conserv 14:31–35 (in Chinese)

Yu XX, Zhao YT, Cheng GW (2002) Distribution patterns and hydrological effects of moss and litter in *Abies fabri* forests on eastern slope of the Gongga Mountain. J Beijing Forestry Univ 24:21–25

Zhang JC, Hu HB (1996) Water and soil loss in China. Chinese Forestry Press, Peking, China (in Chinese)

Zhang JC, Hu HB, Ruan HH, Fang YM (1999) The present situation and control measures of soil and water loss in the Yangtze River Valley. J Nanjing Forestry Univ 2:18–22 (in Chinese)

Zhang JC, Zhuang JY, Nakamura H, Cheng P, Fu J (2004) Suspended sediment discharges from various land uses in the Shangshe catchment in the Mountains, China. Ninth international symposium on river sedimentation, Yichang, China, pp 2426–2431

Zhuang JY, Nakamura H, Zhang JC, Cheng P, Fu J (2004) GIS-based semi-distributed sediment discharge model with a focus on the influence of land use in the Shangshe catchment. J Jpn Soc Erosion Controlling Eng 57:14–22

Chapter 6
Spatial Variability of Soil Erodibility (*K* Factor) at a Catchment Scale in Nanjing, China

Abstract We conducted a study to examine the spatial variability and vertical characteristics of soil erodibility (the *K* factor) and its relationship to vegetation types, based on a case of study on Dengxia catchment, Jiangsu Province, China. Using traditional statistical and geostatistical methods, the *K* factor was calculated by the EPIC model. The results showed the following: (1) the *K* value on the studied area was highly spatially variable, with a range from 0.15 to 0.50, a mean of 0.13, and a coefficient of variance of 22.11%; (2) the overall spatial distribution of the *K* values presented a clear belt shape and was higher in the northwest than in the southeast, with several high *K*-value centers clustered in the central and southern parts of the region. The results showed that the *K* factor of the areas covered by forest, mainly in the northern part, was greater than that of farmland and human residential areas in the central and southern part of the region. (3) The vertical (soil depth) variability of the *K* value by vegetation type followed the order *K*(0–20 cm) < *K* (20–40 cm) < *K* (40–60 cm) overall in the region, while farmland followed a different order, *K* (0–20 cm) < *K* (40–60 cm) < *K* (20–40 cm). It was shown that the *K* factor in the topsoil (0–20 cm) was the greatest, except for farmland. For *K* factor of different vegetation types in the topsoil (0–20 cm), the order was as follows: fallow land > tea garden land > farmland > grassland > broadleaf forestland > bamboo shrub land > coniferous forestland > bamboo forestland.

6.1 Introduction

Soil erodibility, as an index for evaluating the difficulty degree of separation, erosion, and movement of soil by the erosive force of rainfall, is an internal factor involved in soil loss (Lal, 1991; Bo and Li, 1995) and is usually measured by a *K* value. Soil scientists in and outside of China have conducted a large number of studies on the *K* factor. In 1965, Wischmeier and Smith proposed the universal soil loss equation (USLE), based on substantial field data, in which *K* factor was included, and they established for the first time a quantitative relationship between the *K* factor and the amount of soil loss. In 1971, based on the measured correlation between the soil erodibility factor and soil properties, Wischmeier and Smith established a nomograph of the relationship between *K* factor and soil texture, soil organic matter, soil structure, and soil permeability. Williams and Sharply (1990) developed a formula for calculation of the *K* factor in EPIC (erosion productivity impact calculator), relating it only to soil sand, silt, clay, and organic carbon content, thus

J.C. Zhang et al., *Theory and Practice of Soil Loss Control in Eastern China*,
DOI 10.1007/978-1-4419-9679-4_6,

simplifying the calculation of the *K* value. Recently, Chinese soil scientists have also made numerous studies of the *K* factor, including reviews on soil erodibility indices and measuring methods, calculation of the *K* factor of studied areas using different models, and analysis of its spatial distribution characteristics. They also established a revised nomograph (Yang, 1999; Liang and Shi, 1999; Zhang et al., 2001; Bo et al., 2002; Jiang et al., 2004; Song et al., 2006) for the *K* factor of a studied area based on a small-scale measurement of slopes of farmland soil, etc. It can be seen that the *K* factor is an important research topic in the soil and water conservation disciplines; research on the *K* factor has great significance for helping to recognize the mechanisms of soil erodibility, quantitatively estimating soil erosion amounts and finding reasonable means for preventing soil and water loss.

Studies have indicated that the *K* factor is less influenced by soil chemical property than by physical properties that show conspicuous spatial variability even under the same soil conditions. However, previous studies mainly calculated *K* values based on soil type, ignoring their spatial variability (Yang et al., 2006). Guided by geostatistical principles, this chapter focuses on spatial variability characteristics of *K* factor at a catchment scale. The purpose is to better understand soil erodibility spatial distribution characteristics at a catchment scale, providing a scientific basis for *K* factor prediction and soil and water conservation as part of a comprehensive treatment of catchments.

6.2 Materials and Method

6.2.1 General Situation of the Studied Area

Dengxia catchment, China, an area of 32.16 km^2, is situated about 20 km to the southeast of Nanjing City and 20 km to the west of Jurong City, with the NingJu highway running across it. Its elevation ranges from higher in the northwest and lower in the southeast, the highest locale being Mount Tangmu, 282 m high, and the lowest locale being 13 m above sea level. The Huanglongnian reservoir, which belongs to Jurong water system, has a total storage capacity of 22 trillion m^3 with a catchment area of 4.7 km^2. The soil types are mainly yellow brown and red soil, with the greatest soil depth at which freezing occurs being 0.09 m. Belonging to northern subtropical monsoon climate region of China, Dengxia catchment boasts four distinctive seasons, plentiful rainfall, and sufficient light energy resources, with a mean annual temperature at 15.7°C. It rains annually for 117 days with a mean annual rainfall of 1106.5 mm. The best season of a year in regard to these resources is autumn, from September to November. The middle and southern parts of the catchment are mainly farming and residential areas. Most of the forests are located in the northern mountain area, the main vegetation types being Chinese fir (*Cunninghamia lanceolata*), pine (*Pinus massoniana*), German oak (*Quercus acutissima*), white oak (*Quercus fabri Hance*), Zhejiang greyblue spicebush (*Lindera chienii*), oil tea (*Camellia sinensis*), etc.

6.2.2 Soil Sampling

In early July of 2007, we set up sampling sites using a 500-m × 500-m grid and collected soil samples in the study area, exploring the spatial variability characteristics of *K* factor at the catchment scale. To obtain information on the factors of soil texture, terrain, vegetation type, etc., soil sampling was performed at grid intersections. Within a 20 m radius of each intersection point, 3–5 small samples (0–20 cm soil layer) were randomly collected and then integrated into one soil sample. Altogether there were 143 soil samples. GPSMAP 76 was used to record positions of grid intersections (Fig. 6.1). Additionally, we chose typical sample plots (Table 6.1) with different vegetation types (tea garden, coniferous forest, broadleaf forest, grassland, shrub community, bamboo forest, farmland, and fallow land, all together eight types), dug 16 soil profiles in different layers (0–20, 20–40, and 40–60 cm), and collected soil samples in "S" shape on soil profiles. Soil sampling of each vegetation type was replicated for examining the influence of vegetation type on the *K* factor. Collected samples were brought to the lab, naturally air-dried, ground, and screened with a 2-mm sieve for later use. The measured soil properties included bulk density,

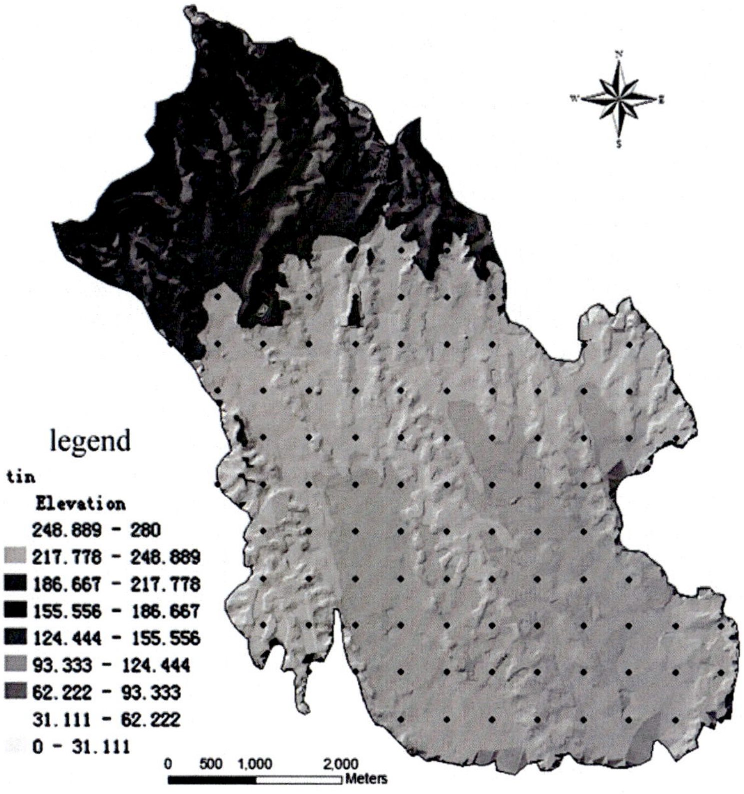

Fig. 6.1 The map of soil sampling sites

Table 6.1 Characteristics of sample plots by different vegetation types

Vegetation type	Stand type	Age (a)	Stand composition	Coverage (%)	Slope (°)	Aspect
S1	II	30	Tea	60	8	SE
S2	II	21	Chinese fir, masson pine	55	12	N
S3	I	24	White oak, Chinese oak, *Lindera glauca*	50	15	SE
S4	I	–	*Carex*, hispid arthraxon herb	90	7	SW
S5	III	8	*Rhus chinensis*	70	9	N
S6	II	13	Bamboo	60	10	NW
S7	–	–	Cotton, sweet potato	65	6	SE
S8	–	–	–	–	5	W

S1, tea garden; S2, coniferous forest; S3, broadleaf forest; S4, grassland; S5, shrub community; S6, bamboo forest; S7, farmland, cotton, and sweet potato; S8, fallow land. I, natural forest; II, plantation forest; III, secondary forest

pH value, mechanical composition, and organic carbon. Soil bulk density was measured by the cutting ring method, pH was measured by potential determination, soil mechanical composition by the hydrometer method (grain size fractionated: 2–0.05, 0.05–0.002, and < 0.002 mm), and soil organic carbon content by the concentrated sulfuric acid–potassium dichromate oxidation process.

6.2.3 Research Methods

6.2.3.1 Data Information

Research data include the 1:10,000 topographic map (2006) of Chunhua Town in Jiangning district of Nanjing, the land-use map based on the field investigation in April 2007, and catchment boundary data in Nanjing Soil and Water Conservation Geographical System (NjSwcGIS). The topographic map is in AutoCad format, while the catchment boundary coordinate is in shape vector format.

6.2.3.2 Calculation of the *K* factor

Calculation methods for the *K* value mainly include direct determination, the nomograph method, and the formula method. Direct determination of the *K* value is considered as the most appropriate to test the sensitivity of practical field soil to eroding force, but it takes a long time and is expensive. The nomograph method requires many parameters, especially accurate knowledge of soil structure grade, which is difficult to access, and the rather trivial soil permeability grade (Bo and Li, 1995), while the formula method is comparatively simple (Wang, 1999; Yang et al., 2006). As far as calculation of *K* value by the American universal soil loss equation (USLE), and its revised edition RUSLE, is concerned, just like the nomograph method, it is hard to determine the necessary indices. Therefore, this study employs the estimation method for the K factor developed from EPIC (erosion productivity impact calculator) by Williams and Sharply (1990). The calculation formula is as follows:

$$K = \left\{0.2 + 0.3\exp\left[-0.0256\mathrm{SAN}\left(1 - \mathrm{SIL}/100\right)\right]\right\} \bullet \left(\frac{\mathrm{SIL}}{\mathrm{SIA} + \mathrm{SIL}}\right)^{0.3}$$
$$\bullet \left[1.0 - \frac{0.25C}{C + \exp\left(3.72 - 2.95C\right)}\right] \bullet \left[1.0 - \frac{0.7SN_1}{SN_1 + \exp\left(-5.51 + 22.9SN_1\right)}\right] \quad (6.1)$$

where

SAN is the sand content (%);
SIL is the silt content (%);
SLA is the clay content (%);
C is the organic carbon content (%); and
SN_1 = 1–SAN/100.

Based on the laboratory-measured practical soil mechanical composition and organic carbon content, formula (6.1) was used to calculate the K value at a catchment scale.

6.2.3.3 Geostatistical Method

The geostatistical method is a spatial analysis method developed from traditional statistics. It can not only effectively demonstrate spatial variability characteristics of attribute variables but can also effectively explain the influence of spatial patterns on ecological processes and functions.

Semi-variance Function Analysis Method

Geo-statistics is a data method based on regionalized variable theory, with the semi-variance function as the basic tool (Wang, 1999). The semi-variance function, which describes spatial variant structures of soil properties, reflects the spatial autocorrelation of observation values at different distances. Its important parameters, such as nuggets, sills, range, and other regionalized variables, are the keys to examining spatial variability of soil properties and bases for spatial distribution estimation. The semi-variance function includes spherical, Gaussian, exponential, linear, and linear-to-sill models. The semi-variance function hypothesizes that a random function has a stable mean, that a variance exists and is limited; and that the value is only related to distance h so that a semi-variance function $\gamma(h)$ can be defined as half of the increment variance of the random function $Z(x)$, that is

$$r(h) = \frac{1}{2}\sigma^2[z(x + h) - z(x)] \quad (6.2)$$

The calculation formula is

$$\gamma(h) = \frac{1}{2N(h)} \sum_{i=1}^{N(h)} [Z(x_i + h) - Z(x_i)] \quad (6.3)$$

where

$\gamma(h)$ represents the semi-variance function;
h represents the spatial interval distance between sampling sites, called the Lag;
$N(h)$ represents the logarithm of observed sampling sites;
$Z(x_i)$ and $Z(x_i + h)$ represents the measured values of the regionalized variable $Z(x)$ at the spatial positions of x_i and $x_i + h$, respectively; and
$\sigma^2[z(x + h) - z(x)]$ represents incremental variance of $Z(x)$.

In this study, Geostatistical Analyst in ArcGIS(ESRI) was used for calculating the semi-variance function analysis of 143 K values and the kriging method was used for an optimum unbiased estimate. The Lag was 0.625 km, significantly smaller than the range of variance function (7.4083 km), in accordance with the requirements of geostatistical analysis.

Kriging Method

The kriging method is the most common interpolation method in geo-statistics. It is used for optimum unbiased estimate on regionalized variables of non-sampling sites based on the structure of original data and semi-variance function (Lu et al., 1997). The method works as follows: assume there is a particular variable at the position x_0 in a region $Z^*(x_0)$. In the related area around it, there are n measured values $Z(x_i)(i = 1, 2, \ldots, n)$; then a linear combination of n measured values $Z(x_i)$ is used for the estimate $r(h) = \frac{1}{2}\sigma^2[z(x + h) - z(x)]$, that is

$$Z^*(x_0) = \sum_{i=1}^{n} \lambda_i Z(x_i) \quad (6.4)$$

where λ_i represents a weighted coefficient related to position $Z(x_i)$. The premium estimate must meet two conditions: (1) unbiased estimate – $E[Z^*(x_0) - Z(x_0)]$ – and (2) the minimum variance – $\sigma^2[Z^*(x_0) - Z(x_0)] = \min$. Based on the Lagrange minimization principle, from the above formulas, a matrix relating λ_i and semi-variance can be derived:

$$\begin{bmatrix} r_{11} & r_{12} & \ldots & r_{1n} & 1 \\ r_{21} & r_{22} & \ldots & r_{2n} & 1 \\ & & \ldots & & \\ r_{n1} & r_{2n} & \ldots & r_{n3} & 1 \\ 1 & 1 & \ldots & 1 & 0 \end{bmatrix} \begin{bmatrix} \lambda_1 \\ \lambda_2 \\ \ldots \\ \lambda_n \\ \varphi \end{bmatrix} = \begin{bmatrix} r_{10} \\ r_{20} \\ \ldots \\ r_{n0} \\ 1 \end{bmatrix} \quad (6.5)$$

In the formula, the distance between r_{ij}–x_i and x_j represents the semi-variance of $|x_i - x_j|$ and φ is a Lagrange operator. After working out λ_i and φ from the above equations, the premium estimate $Z^*(x_0)$ at the point x_0 can be derived from Eq. (6.5).

6.3 Results and Analysis

6.3.1 Descriptive Statistical Analysis of the K Factor

6.3.1.1 Descriptive Statistical Features

Analyzed by classical statistical methods, the statistical features of *K* factor in the studied area are listed in Table 6.2. It can be seen from Table 6.2 that the range of *K* factor was 0.1498–0.4981, the maximum being 3.3 times of the minimum, with a rather large amplitude. The mean of *K* value was 0.3316 and median was 0.3295. They are close to each other, which implies that the *K* factor of the study area was well distributed and uninfluenced by outliers. The coefficient of variation, the percentage of standard deviation to mean, reflects the dispersion of samples. It is graded as follows: small variability CV < 10%, medium variability CV = 10–100%, and strong variability CV > 100%. Therefore, the *K* factor of the studied area demonstrated moderate spatial variability, which was caused by a variety of factors, such as soil texture, terrain, vegetation, and farming methods.

Table 6.2 Statistical features of *K* factor

Samples	Min.	Max.	Mean	Median	Std. dev.	CV (%)	Skew	Kurtosis
143	0.1498	0.4981	0.3316	0.3295	0.0733	22.11	–0.117	–0.076

6.3.1.2 Normal Distribution Test

Classical statistical analysis can only summarize general features of *K* factor of the studied area rather than local variation features; it only reflects whole samples to a certain degree, rather than being a quantitative description of the randomness and structure, independence, and correlation. Therefore, the above results cannot reflect the spatial relationship and variation features of the *K* factor; geostatistical methods must be further employed for spatial variability analysis of the *K* factor.

The normal distribution of data is the basis for spatial analysis of the *K* factor with geostatistical methods (Lu et al., 1997; Lei et al., 1985; Wang, 1999). The frequency distribution of the *K* factor (Fig. 6.2) can be derived by using the frequency distribution statistical function called Frequencies in SPSS 15.0. Figure 6.2 shows that the frequency distribution of the *K* factor looked like a bell curve, basically in accordance with the normal distribution. To further identify its distribution type, the Kolmogorov–Smirnov test was employed, with a goodness of fit of 0.051. Therefore, the *K* factor was in accordance with the normal distribution and the assumed conditions of geostatistical analysis.

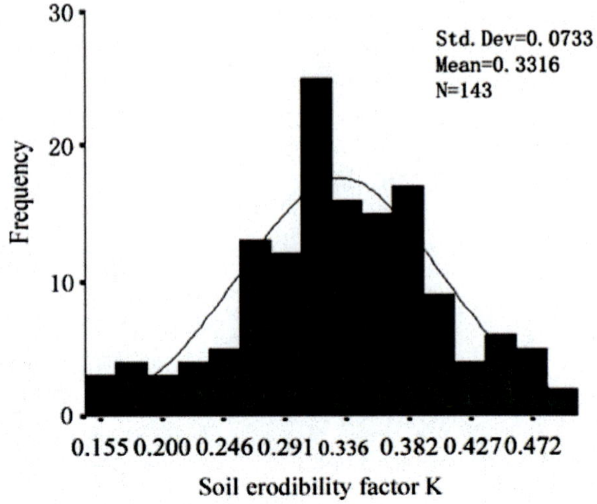

Fig. 6.2 Frequency distribution of the *K* factor

6.3.2 Semi-variance Function Analysis of the K Factor

The semi-variance function is the theoretical basis for analyzing the *K* factor using geo-statistics. It has three important functions, namely nugget, range, and sill. Range reflects spatial variability of the *K* factor; *K* values outside of the range are spatially independent, while *K* values within the range are correlated. Nugget is caused by random factors, including measurement errors and variation in soil properties (fertilization, tillage practices, cropping system, etc.) at the minimum sampling distance. Sill reflects the influence of structural factors (soil type, parent material, terrain, climate, etc.) on regionalized variables. Spatial correlation of the *K* factor can be ranked according to the ratio of nugget to sill, or $C_0/(C_0 + C)$ (Zheng et al., 2003). High ratios imply strong spatial variability caused by random factors, while low ratios imply strong spatial variability caused by structural factors. When $C_0/(C_0 + C) < 25\%$, variables show strong spatial correlation; when $C_0/(C_0 + C)$ is within 25–75%, variables show medium spatial correlation; when $C_0/(C_0 + C) > 75\%$, variables show weak spatial correlation (Cambardella et al., 1994). The choice of the semi-variance function model is the key to spatial variability analysis of the *K* factor. In this study, the Geostatistical Analyst model in ArcGIS was used to choose the semi-variance function theoretical model (Table 6.3). We calculated the experimental variation function value of the *K* factor by formula (6.3), compared $C_0/(C_0+C)$, lag, and range of different models, and chose an index model as semi-variance function theoretical model for the *K* factor (Fig. 6.3).

Table 6.3 and Fig. 6.3 show that when the semi-variance function theoretical model of the *K* factor was an index model, nugget was 0.0010418, sill was 0.0069231, $C_0/(C_0 + C)$, therefore, was 15.05%, or less than 25%, which demonstrated strong spatial correlation within range, and structural factors were the main

Table 6.3 Theoretical semi-variogram models of *K* factor and related parameters

Item	Theory model	Nugget (C_0)	Partial sill	Sill (C_0+C)	$C_0/(C_0+C)$	Lag (km)	Range (km)
K value	Spherical model	0.0016654	0.0057175	0.0073829	0.2256	0.625	7.4083
	Exponential model	0.0010418	0.0058813	0.0069231	0.1505	0.625	7.4083
	Gaussian model	0.0023385	0.0063603	0.0086988	0.2688	0.625	7.4083

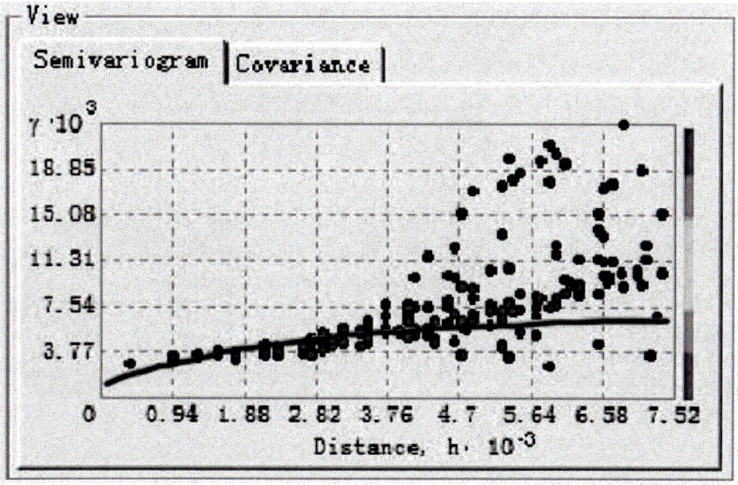

Fig. 6.3 Semi-variogram of *K* factor

factors of spatial variability. With lag of 0.625 km and range of 7.4083 km, range was much larger than lag, which indicated good spatial correlation at a catchment scale. Accurate results could be derived with a kriging method.

6.3.3 Spatial Variation Features of K Factor

We chose kriging of Geostatistical Analyst in ArcGIS, set the pixel size of output grid as 10 m × 10 m, acquired spatial relationships while using the semi-variation function and cross-validation function for fitting, and finally generated a spatial variability map of the *K* factor (see Fig. 6.4).

Figure 6.4 shows that the spatial distribution of the *K* value was conspicuously belt shaped and was higher in the southeast than in the northwest, with several high *K*-value centers embedded in the central and southern parts of the region and the highest being in the southwestern part of the region. This spatial distribution feature of the *K* value was mainly related to the strong spatial variability caused by soil

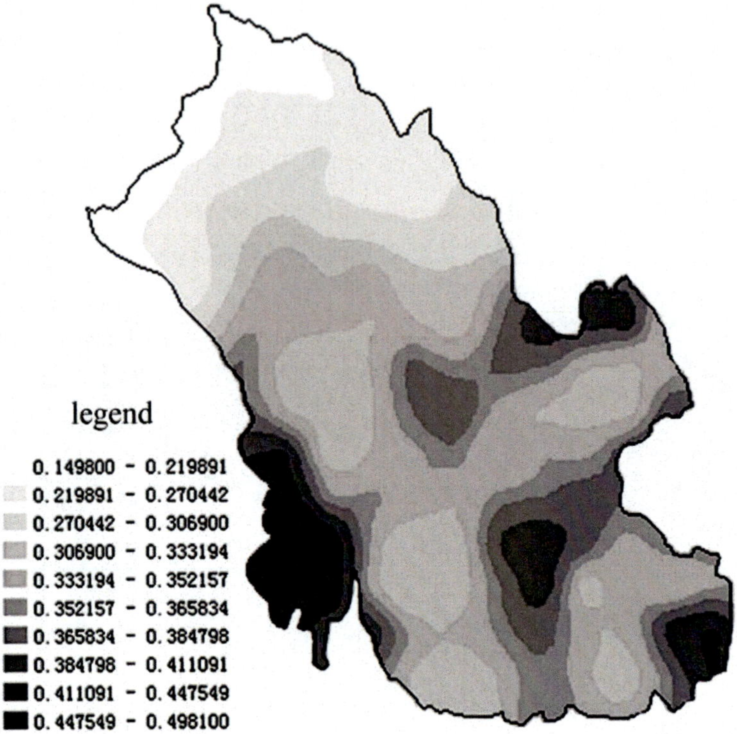

Fig. 6.4 Spatial variability map of *K* factor

structural factors (soil type, parent material, terrain, climate, etc.) and the increased spatial correlation of soil erodibility caused by random factors (fertilization, tillage practices, cropping system, etc.). The elevation of study area is high in the northwest and low in the southeast, high in the east and west, and low in the middle (Fig. 6.1). The northern part is forest covered, while the middle and southern parts are farming and residential areas. The diversity of the underlying surfaces of different regions in the area and different degrees of human disturbance were the main random factors in the distribution feature of *K* value. In the areas covered by forest, the main factor influencing the *K* value was vegetation type. The growth and the distribution of roots of different plant types were different, as was the accumulation of soil organic matter, which influenced soil anti-erodibility. In farming and residential areas of the middle and south, low terrain, long-time human cropping, overexploitation, and overuse of land led to accelerated decomposition and low accumulation of organic matter, severely interrupting soil particle composition; therefore the *K* value increased and soil anti-erodibility weakened. Obviously, the soil erosion resistance of the forested area in the catchment was strong, soil erosion resistance of the farming and residential areas of the middle and south was weak, while soil erosion resistance of some local areas with high soil erodibility *K* values was the weakest.

6.3.4 *Vertical Variability Characteristics of K Factor by Different Vegetation Types*

The cropping system is an important factor in spatial variability of soil property. In order to examine the influence of different vegetation types on the *K* factor, we chose eight types of typical sample plots – tea garden, coniferous forest, broadleaf forest, grassland, shrub community, bamboo forest, farmland and fallow land, and dug soil profiles in different layers (0–20, 20–40, 40–60 cm) for the study of vertical variability characteristics of the *K* factor (see Fig. 6.5).

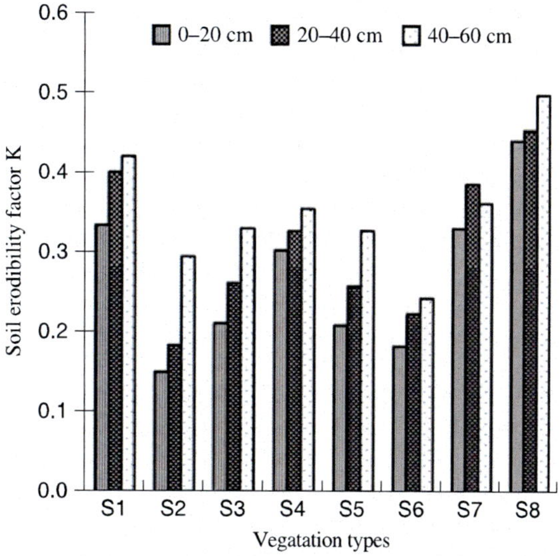

Fig. 6.5 Vertical variability characteristics of the *K* factor for different vegetation types

It can be seen from Fig. 6.5 that the vertical variability characteristics of the *K* values for the vegetation types as a whole followed the order of $K_{0\text{–}20\ \mathrm{cm}} < K_{20\text{–}40\ \mathrm{cm}} < K_{40\text{–}60\ \mathrm{cm}}$ in the area, while farmland followed the order of $K_{0\text{–}20\ \mathrm{cm}} < K_{40\text{–}60\ \mathrm{cm}} < K_{20\text{–}40\ \mathrm{cm}}$. The figure shows that soil anti-erodibility in the topsoil (0–20 cm) was the strongest, except for farmland, whose soil properties were greatly changed by fertilization, tillage practices, and cropping restructuring. For the *K* factors of different vegetation types in the topsoil (0–20 cm), the order was as follows: fallow land > tea garden land > farmland > grassland > broadleaf forestland > bamboo shrub land > coniferous forestland > bamboo forestland. This demonstrates that vegetation could improve soil properties, increase soil erosion resistance, and decrease soil erodibility. From formula (6.1), it can be seen that the *K* factor is related to soil particle composition and soil organic carbon content. Vegetation can improve soil properties, increase organic carbon content, reduce soil erodibility, and increase soil anti-erodibility. In view of the distributions of roots of different vegetation types in soil vertical profiles, the diversity of sampling sites' soil parent matter (terrain), and the degree of human disturbance, further studies are needed to explore reasons for the differences in variation ranges of the *K* factor for different vegetation types.

6.4 Conclusions

(1) It was shown to be highly feasible to use traditional statistical and geostatistical methods, to collect soil samples by the grid method, to calculate the *K* factor using the EPIC model, and to generate a spatial variability graph of the *K* factor by using the kriging method in ArcGIS. The *K* value in the study area showed significant spatial variability, with a range of values from 0.1498 to 0.4981, a mean of 0.3316, and a coefficient of variance of 22.11%. Therefore, spatial variability of *K* factor should be considered for quantitative monitoring of soil erosion and evaluation of soil and water loss at the catchment scale.

(2) The spatial distribution of *K* values was conspicuously belt shaped and it was higher in the southeast than in the northwest, with several high *K*-value centers embedded in the middle and southern parts of the region, with the highest values being in the southwestern part of the region. It can be seen that soil erosion resistance of the forested area of the catchment was strong; soil erosion resistance of the agricultural and residential areas of the middle and south part was weak; and soil erosion resistance of some local areas with high *K* values was the weakest.

(3) The vertical variability characteristics of the *K* values for different vegetation types followed the order of $K_{0\text{–}20\ \mathrm{cm}} < K_{20\text{–}40\ \mathrm{cm}} < K_{40\text{–}60\ \mathrm{cm}}$ as a whole in the region, while in farmland they followed the order of $K_{0\text{–}20\ \mathrm{cm}} < K_{40\text{–}60\ \mathrm{cm}} < K_{20\text{–}40\ \mathrm{cm}}$. The study showed that soil erosion resistance was strongest in the topsoil (0–20 cm), except for farmland. For the *K* factors of different vegetation types in the topsoil (0–20 cm), the order was as follows: fallow land > tea garden land > farmland > grassland > broadleaf forestland > bamboo shrub land > coniferous forestland > bamboo forestland.

References

Bo ZH, Li QY (1995) Preliminary study on the methods of soil erodibility value mapping. Rural Eco-Environ 11(1):5–9

Bo ZH, Yang ZL, Bo YH et al (2002) Soil erodibility (*K*) value and its application in Taihu lake catchment. Acta Pedol Sin 39(3):296–300

Cambardella CA, Moorman TB, Novak JM (1994) Field scale variability of soil properties in Central Iowa soils. Soil Sci Soc Am J 58:1501–1511

Jiang XS, Pan JJ, Yang ZL et al (2004) Methods of calculating and mapping soil erodibility *K*. Soils 36(2):177–180

Lal R (1991) Soil and water conservation academy, translation by the publication center of yellow river resources committee. Erodibility and erosion. Science Press, Beijing

Lei ZD, Yang SD, Xu ZR et al (1985) Preliminary study on soil characteristics spatial distribution. J Hydraul Eng 9:10–21

Liang Y, Shi XZ (1999) Soil erodible K in eastern hillyfields of the southern Yangtze River. Res Soil Water Conserv 6(2):47–52

Lu WD, Zhu YL, Sha J et al (1997) SPSS for windows from rudiment to mastery. Electron Industry Press, Beijing

Song Y, Liu LY, Yan P et al (2006) A review of soil erodibility research. Arid Land Geogr 29(1):124–131

Wang ZQ (1999) Geostatistics and its application in the ecology science. Science Press, Beijing

Williams JR, Sharply AN (1990) EPIC-erosion productivity impact calculator: 1. Model documentation. US Department of Agriculture Technical Bulletin

Yang ZS (1999) Soil erodibility factor of sloping cultivated land in the northeast mountain region of Yunnan Province. J Mt Sci 17(sup):10–15

Yang P, Hu XL, Jiang XS et al (2006) Spatial variability of soil erodibility *K* value and influence sampling densities on *K* value accuracy at a scale of small watershed. Bull Soil Water Conserv 26(6):35–39

Zhang KL, Cai YM, Liu BY et al (2001) Evaluation of soil erodibility on the Loess plateau. Acta Ecol Sin 21(20):1686–1695

Zheng YM, Chen H, Chen TB et al (2003) Spatial distribution patterns of Cr and Ni in soils of Beijing. Quaternary Sci 23(4):436–445

Chapter 7
Application of a GIS-Based Revised FER-USLE Model in the Shangshe Catchment

Abstract To support policy decisions, the greatest need is for models that can produce robust results using readily available data. The ER-USLE model was tested at the sub-catchment scale in a sub-catchment of cultivated land by comparing observed specific sediment discharge (*SSD*) at the outlet of sub-catchment with predicted *SSD* using ER-USLE model. The FUSLE model was tested from the USLE plot to the sub-catchment scale in sub-catchments of pine forest, Chinese fir forest, and tea garden by comparing observed *SSD* with predicted *SSD* using FUSLE model. Integrating the ER-USLE model for soil loss prediction in cultivated land and the FUSLE model for soil loss prediction in forest, the FER-USLE model was developed for soil loss prediction in both cultivated land and forest. The FER-USLE model was tested in the Shangshe catchment by comparing its predictions with observed *SSD* at the river outlet, showing robust results. Results show that it is useful for prediction of annual soil loss for various types of land use in the Dabie Mountains, China. FER-USLE model is available to many users and policy makers in the Dabie Mountains, China.

7.1 Introduction

Erosion can be examined over a wide range of spatial scales. This includes small sample plots for scientific study, the field scale that is of interest to individual farmers, the catchment scale for community level issues, and regional and national scales for policy makers (Kirkby et al., 1996). In the policy arena, the impacts of human activity on soil erosion have become a major concern. Erosion occurs from both extreme events and lower magnitude events. The extreme events engender major soil erosion, whereas the lower magnitude events reduce agricultural productivity and increase water pollution. An important challenge in soil erosion modeling, which has become even more important with the increasing use of models linked to GIS, is the mismatch between the small spatial and temporal scales of data collection and model conceptualization, and the large spatial and temporal scales of most intended uses of models (Renschler and Harbor, 2002). Scientists are well aware of such upscaling problems. Robust models are expected to deal explicitly with variability, as well as with issues of how data on erosional processes at one scale can be extrapolated to processes that operate at other scales. Because landscapes are spatially heterogeneous areas, structure, function, and temporal change of landscapes are scale dependent themselves (Turner, 1989). Therefore, each regionalization method

J.C. Zhang et al., *Theory and Practice of Soil Loss Control in Eastern China*,
DOI 10.1007/978-1-4419-9679-4_7,

must be developed and validated to fulfill requirements for a specific purpose at a specific scale.

The ER-USLE model was developed from USLE with addition of an effective rain erosivity factor. It was modified (FUSLE) for application to a sub-catchment of Chinese fir forest in Chapter 5. In this study, ER-USLE was integrated with the FUSLE model for soil loss prediction in all types of land use in the Shangshe catchment.

7.2 Study Area

This study was carried out in the Shangshe catchment (34°32′20″N,116°50′12″E) of the Dabie Mountains in Yuexi prefecture of Anhui Province, China (Fig. 2.1). Details of the Shangshe catchment were introduced in Chapter 2.

7.3 Materials and Methods

7.3.1 Field Observations of Soil Loss at the Micro-plot Scale, the USLE-Plot Scale, the Sub-catchment Scale, and the Catchment Scale

Field observations of soil loss at the micro-plot scale were applied to the sub-catchments of cultivated land and Chinese fir forest, in the year 2000. These applications were introduced in detail in Chapters 4 and 5, respectively. Similar field observations of soil loss at the micro-plot scale were carried out in the sub-catchments of the tea garden and the pine forest, also in the year 2000.

Field observations of soil loss at the sub-catchment and USLE-plot scales of cultivated land, Chinese fir forest, tea garden, pine forest, and the outlet of the Shangshe catchment were carried out over the period from year 2000 to 2008. This was introduced in detail in Chapter 2.

7.3.2 Field Observations of Litter Coverage and Terrace Conditions

Based on Fig. 2.2, the litter coverage and terrace condition were measured by eye through field observations in the Shangshe catchment.

7.3.3 FER-USLE Model

ER-USLE model was developed from the USLE model using an effective rain erosivity factor in the sub-catchments of cultivated land. The USLE model was modified for use in the sub-catchments of Chinese fir forest (FUSLE). In this Chapter, the ER-USLE and FUSLE models are integrated for soil loss prediction in all types of lands.

Equation (7.1) is obtained for soil loss prediction for single events:

$$A_e = R_e LSKPP_s C_l C_s l_t \tag{7.1}$$

where

A_e is the *SSD* (specific sediment discharge) for single events (t km^{-2} or g m^{-2}) (at the micro-plot and USLE-plot scale, the units are (g m^{-2}), while at the sub-catchment and catchment scales, the units are converted into (t km^{-2}) in this study);
R_e is the effective rain erosivity factor (J mm m^{-2} h^{-1});
L is the slope length factor, S is the slope factor;
P_s is the single operation factor in cultivated land;
P is the support-practice factor, which affects soil loss over a long time period;
K is the soil erodibility factor (g J^{-1} mm^{-1} h);
C_l is the factor of land use; C_s is the seasonal coverage factor; and l_t is the litter factor.

Equation (7.2) is obtained for annual or seasonal loss prediction based on summation over single precipitation events:

$$\sum_{i}^{n} A_{ei} = \sum_{i=1}^{n} R_{e_i} LSKPP_{si} C_l C_{si} \tag{7.2}$$

where

$\sum_{i}^{n} A_{ei}$ is the annual or seasonal *SSD* (t km^{-2} or g m^{-2});
n is the total number of the precipitation events, and the other factors are the same as those in Eq. (7.1).

Because this model is developed from the ER-USLE model and the FUSLE model, it is named the FER-USLE model.

7.4 Calculations of Factors in FER-USLE Model

7.4.1 R_e and K Factors

The calculated R_e for the 3 years from 2000 to 2002 in Chapter 4 was utilized in the Shangshe catchments. The K factor was calculated in Chapter 4 for cultivated land and in Chapter 5 for Chinese fir forest. The same method was used to calculate the K factor in pine forest and tea garden. In addition, the calculated value of the K factor for each type of land use was assumed to be the same for all other areas of land of the same type, and applied to corresponding land types here.

7.4.2 LS Factor

The topographic slope was derived from the digital elevation model (DEM) of the Shangshe catchment of cultivated land at the 5 m × 5 m scale of resolution (Fig. 3.3) and is shown in Fig. 7.1. In both the USLE and RUSLE models, slope length is defined as the horizontal distance from the origin of overland flow to the point where either the slope gradient decreases to a point at which deposition begins or runoff becomes concentrated in a defined channel (Wischmeier and Smith, 1978; Renard et al., 1997). In the present study, the L factor was determined using the following method.

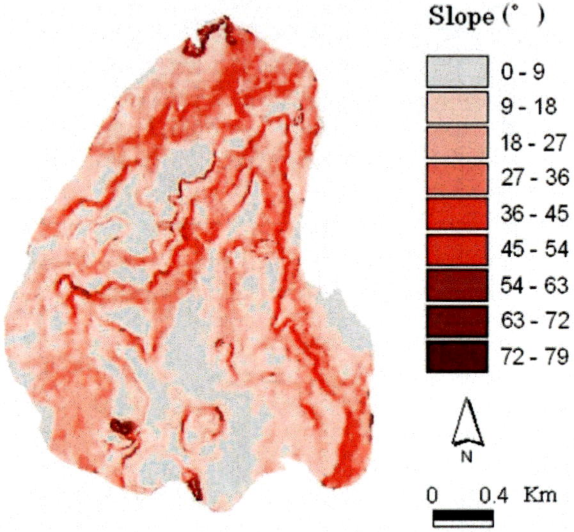

Fig. 7.1 Slope distribution of the Shangshe catchment

First, the Shangshe catchment was divided into 87 sub-catchments with the DEM using Arc View software under a hydrological model. In each sub-catchment, the longest flow path was calculated (Fig. 7.2). Then, each sub-catchment was divided into two sub-catchments along the longest flow path. The average slope length in each sub-catchment is calculated using Eq. (7.3), and is shown in Fig. 7.3:

$$L_{slope} = S_A / L_{stream} \tag{7.3}$$

where

S_A is the area of a sub-catchment (m^2);
L_{stream} is the length of the longest flow path in the sub-catchment (m); and
L_{slope} is the average length of slope in the sub-catchment (m).

The *LS* factor was calculated according to (4.10); the *L*, *S*, and *LS* factors are shown in Figs. 7.4, 7.5, and 7.6, respectively.

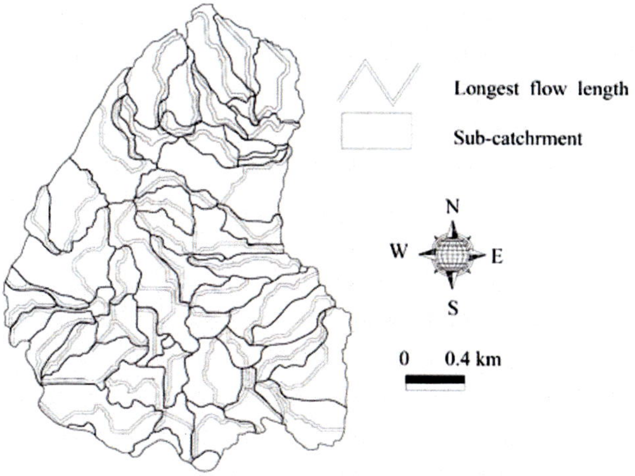

Fig. 7.2 Sub-catchments and longest flow path in each subcatchment

Fig. 7.3 Slope lengths in the Shangshe catchment

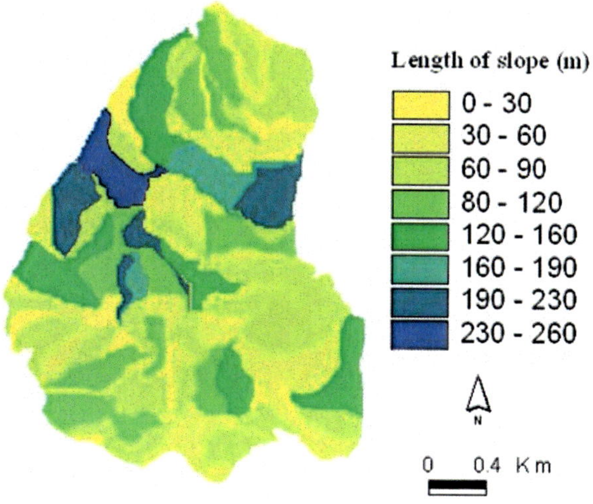

7.4.3 P Factor

The main soil conservation techniques used in the Dabie Mountains are contour tillage, contour farming, and level terraces. The support-practice factor (*P*) is the soil loss ratio for a specific support practice to the corresponding soil loss for up-and-down slope tillage (Renard et al., 1997).

The *P* value in the sub-catchment of cultivated land was calculated in Chapter 4. It is equal to 0.11. The *P* values in the other areas of cultivated land and tea garden were determined through field observation: the *P* values were divided into 10 classes from 0 to 1. The method to determine the *P* value is listed in Table 7.1.

Fig. 7.4 *L* factor in the Shangshe catchment

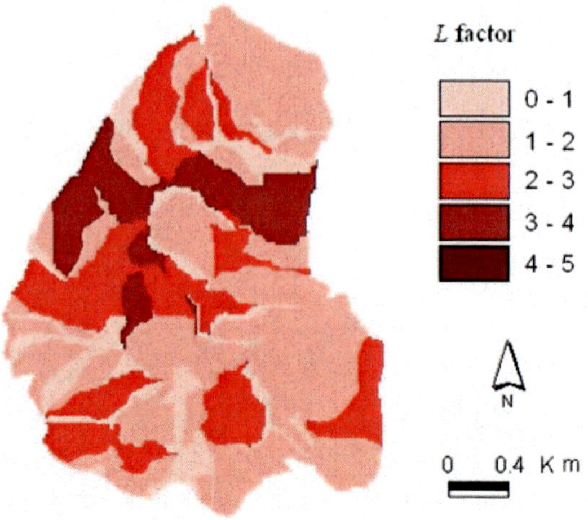

Fig. 7.5 *S* factor in the Shangshe catchment

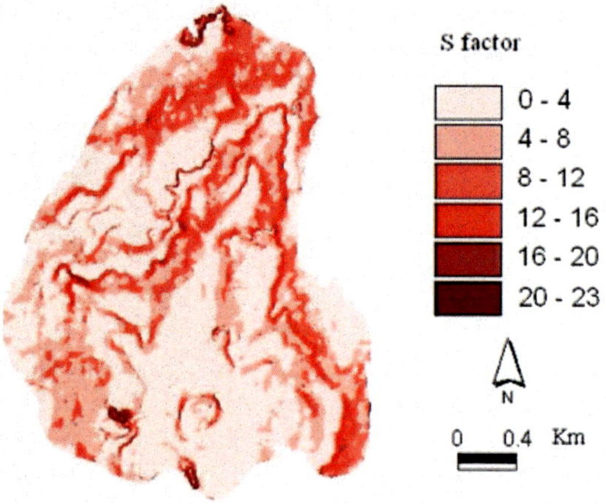

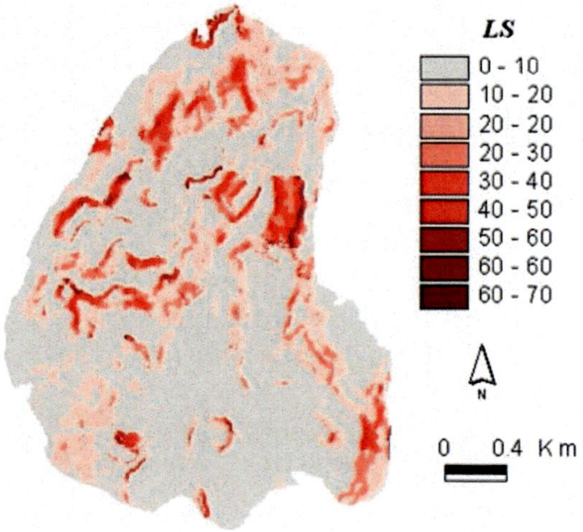

Fig. 7.6 Spatial distribution of the *LS* factor

Table 7.1 *P* value and the support-practice measures in the cultivated land and tea garden in the Shangshe catchment

Measures	*P* value
Tillage along the slope without terrace in cultivated land (S_{NTC})	1.0
Paddy field and water level terrace	0.0–0.1
S_{CNT}:S_{TC} = 0:1–1:1	0.1–0.3
S_{TC}:S_{CNT} = 0:1–1:1	0.3–0.5
S_{CNT}:S_{NTC} = 0:1–1:1	0.5–0.9

S_{TC}, tillage along the contour with a terrace of 1–5 m width. This measure is in the sub-catchment of cultivated land and its value was calculated to be 0.11. S_{CNT}, tillage along the contour without terrace. Its value was reported to be 0.5 by yang (2002)

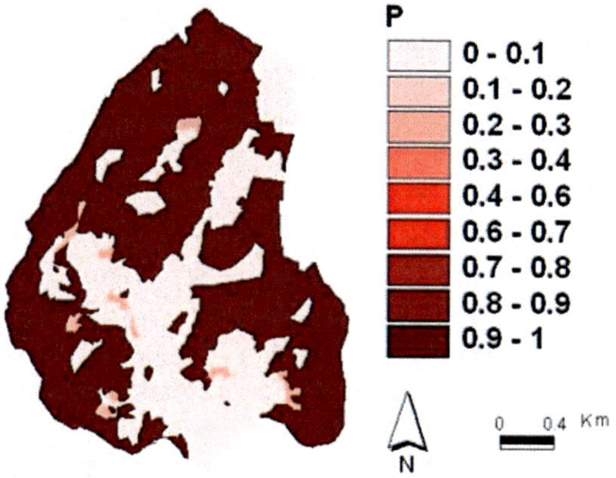

Fig. 7.7 Spatial distribution of the *P* factor

Table 7.1 shows *P* values and the support-practice measures in the Shangshe catchment. The *P* value for the forest area was estimated to be 1, while that for the paddy field was estimated to be 0. The distribution of *P* factors in the Shangshe catchment was estimated and is shown in Fig. 7.7.

7.4.4 P_s and C_s Factors

The effect of daily operations on annual soil loss in cultivated land is represented by P_S. The effect of seasonal coverage on annual soil loss is C_S. In Chapter 4, using Eq. (4.1) to predict annual soil loss without considering the effect of P_S and C_S factors, it was assumed that the predicted seasonal values cannot agree well with the observed values. The ratio of observed seasonal soil loss to corresponding predicted seasonal soil loss using Eq. (4.9) is defined as $P_S C_S$. In ER-USLE the C_S factor reflects the effects of seasonal variation of plant coverage on soil erosion rates. It is suitable for cultivated land and for forestland consisting of deciduous trees. However, in pine and Chinese fir forests, and in tea gardens, where trees or shrubs are evergreen, the coverage does not show seasonal variations. The effects of operations by humans are also negligible in such areas. Therefore, the values of $P_S C_S$ were equal to unity from March to October in the forest and tea garden. Explanations of the C_S and P_S factors in cultivated land were introduced in Chapter 4. However, it was reported in Chapters 4 and 5 that in cultivated land and Chinese fir forest, field observations for the 3 years of 2000-2002 showed no soil losses for the months of November, December, January, and February. Therefore, R_e was calculated from March to October in the 3 years of this study.

7.4.5 C_l Factor

In Chapter 4, the ER-USLE model was proposed and applied in the cultivated land. In Chapter 5 the FUSLE model was developed in the sub-catchments of Chinese fir forest. The C_l factor in each type of land use was calculated from linear regressions of *SSD* with $R_e LS$ based on micro-plot soil loss data in the year 2000 (Zhuang, 2006). The C_l factors in each type of land use are listed in Table 7.2.

Table 7.2 C_l factor in the Shangshe catchment

Land use	C_l
Pine forest	0.041
Tea garden	0.012
Chinese fir forest	0.0124
Cultivated land	0.17

7.4.6 Litter Factor

In Chapter 5, the value of L_i in the sub-catchment of Chinese fir forest was calculated using Eq. (5.3) from observed micro-plot data in the year 2000. This value was used in the other Chinese fir forests in the Shangshe catchment despite the fact that the spatial distribution of thickness of litter is quite different. The value of L_i in the sub-catchment of Chinese fir forest was calculated from observed *SSD* of micro-plots in the year 2000. The same method was used to calculate the values of L_i in the sub-catchments of pine forest and tea garden (Table 7.3). In the cultivated land, the value of L_i is 1.

Table 7.3 Values of L_i in four types of land use in the Shangshe catchment

Land use	L_i
Pine forest	0.014
Tea garden	0.0595
Chinese fir forest	0.0582
Cultivated land and bare land	1

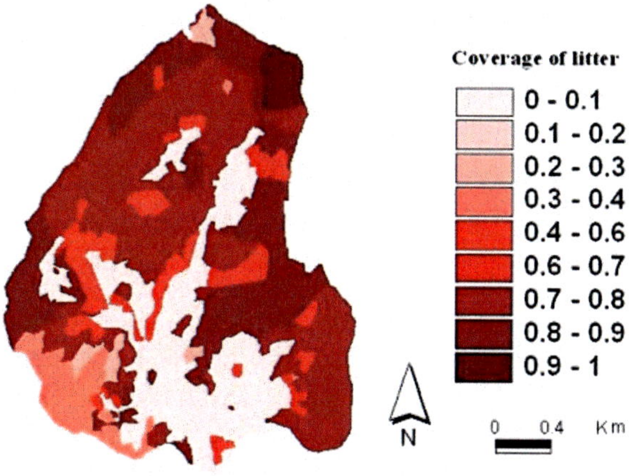

Fig. 7.8 Spatial distribution of coverage of litter on the ground in the Shangshe catchment

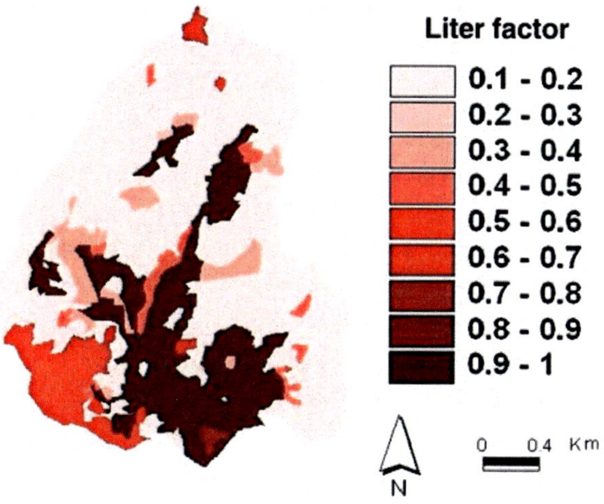

Fig. 7.9 Spatial distribution of litter factor in the Shangshe catchment

Coverage of litter on the ground surface was estimated by eye in field observations. It is shown in Fig. 7.8. The litter factor was calculated using Eq. (5.3) and is shown in Fig. 7.9.

7.5 Results

(1) Comparison between observed values of *SSD* at the river outlet with predicted *SSD* in the Shangshe catchment with the FER-USLE model

With the map layer of *LS* and the other factors introduced above, the predicted soil loss for single rainfall-runoff events in the year 2000 in the Shangshe catchment

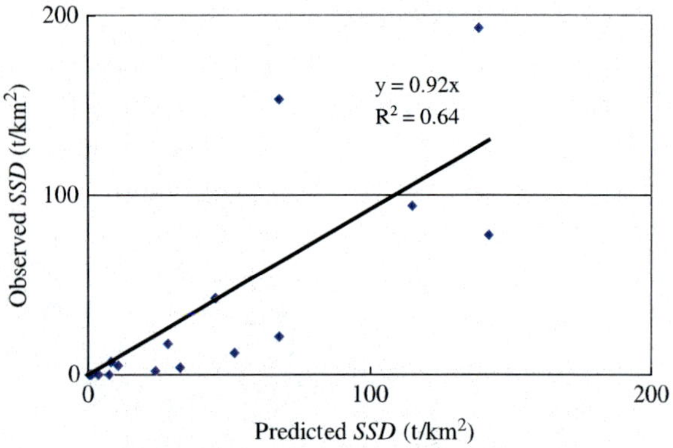

Fig. 7.10 Comparison of observed *SSD* at the river outlet with predicted *SSD* using the FER-USLE model in the year 2000

was calculated using the FER-USLE model. The values were compared with the observed values at the river outlet in the year 2000 (Fig. 7.10). Figure 7.10 shows that the observed values are 0.92 times the predicted values on average. Their R^2 is 0.64.

(2) Prediction of annual soil loss in the Shangshe catchment in 2000

With the map layer of *LS* and the other factors introduced above, the predicted annual soil loss in 2000 was calculated using the FER-USLE model. It is shown in Fig. 7.11. The values of annual *SSD* were distributed between 0 and 300 (t km^{-2}) in the paddy field and forest with high litter coverage. Values of annual *SSD* were distributed between 1000 and 3000 (t km^{-2}) in tea garden and forest with little litter

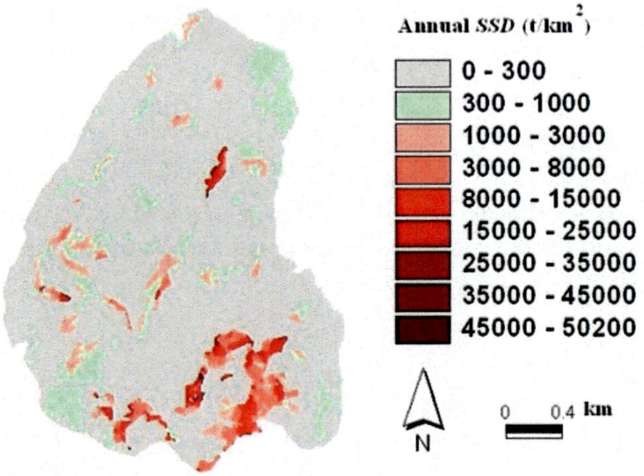

Fig. 7.11 Spatial distribution of predicted annual *SSD* in the year 2000

coverage. The values of annual *SSD* larger than 3000 (t km^2) were distributed mainly in cultivated land.

7.6 Percentage of Predicted *SSD* by Land Use Using the GOIUG and FER-USLE Models

In Chapter 3, the GOIUG (GIS-based observed instantaneous unit graph) model was applied to simulate the *SISSD* (specific instantaneous suspended sediment discharge graph based on the observed *SISSD* graphs from four types of land use. The annual soil loss from each type of land use predicted using GOIUG method is calculated using Eq. (2.11). Figure 7.12 shows that in 2000 the predicted annual sediment discharge at the river outlet with GOIUG is 3901.6 (t), the annual soil loss predicted with FER-USLE in the Shangshe catchment is 3874.2 (t), and the observed *SSD* at the river outlet is 3235.8 (t). Because sedimentation and gully erosion are not considered in this study, this is reasonable. However, Figs. 7.13 and 7.14 show that percentages of predicted soil loss by land use in the Shangshe catchment with the two models agree well in the cultivated land but differ greatly in the forest and tea garden. The GOIUG model assumes that the observed *SISSD* graph can be applied directly in the other areas having the same land use as in the Shangshe catchment. In fact, this is not true. The results simulated by the FER-USLE model suggest that even among the same types of forestland use in the Shangshe catchment, annual soil loss is quite different. The GOIUG model applied the GIS-based unit graph method for temporal soil loss prediction. Based on the observed *SISSD* graphs from various kinds of land use, the geomorphology, plant coverage, and litter factors should also be considered for application at the catchment scale. This is planned for development in future studies.

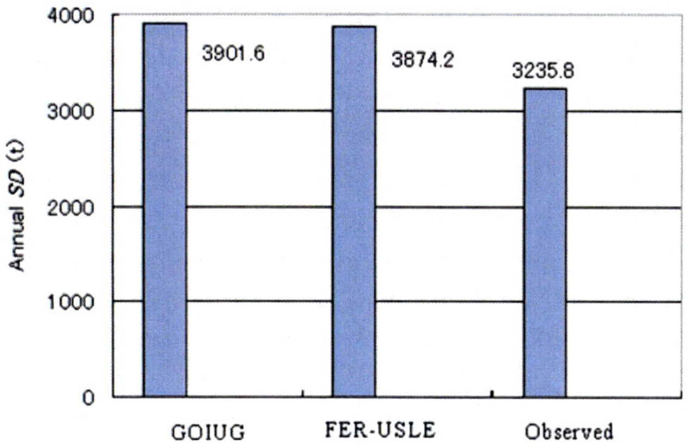

Fig. 7.12 Observed annual *SD* at the river outlet and predicted annual *SD* using the GOIUG and FER-USLE models in the Shangshe catchment

Fig. 7.13 Percentage of predicted annual *SSD* by land use using the FER-USLE model in 2000 in the Shangshe catchment

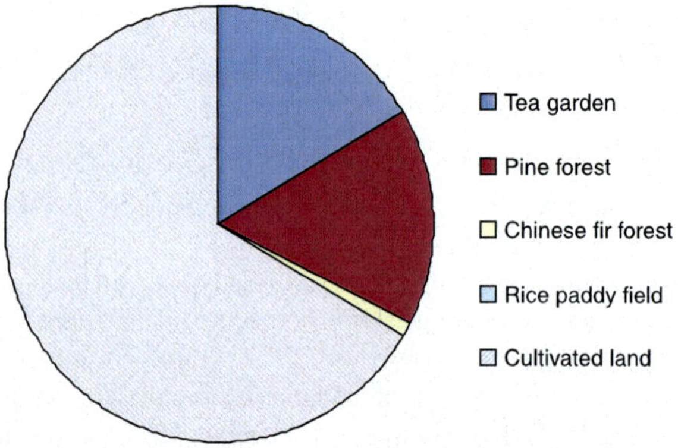

Fig. 7.14 Percentage of predicted annual *SSD* by land use using the GOIUG model in 2000 in the Shangshe catchment

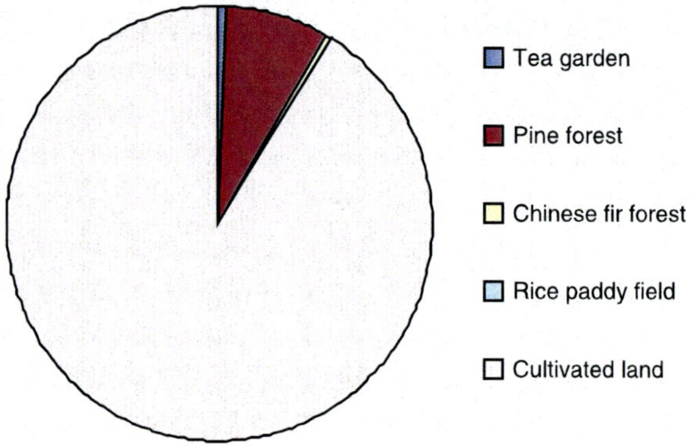

7.7 Conclusions

In this study, field observations of water and sediment discharge were carried out at the micro-plot scale, the USLE-plot scale, the sub-catchment scale, and the catchment scale in the Shangshe catchment, Dabie Mountains, China.

Through analyses of field observation data in the Shangshe catchment in the 3 years of 2000–2008, the following conclusions were reached:

(1) Considering the variation of seasonal coverage of crop and short timescale operations by human beings, C_s and P_s were incorporated into the ER-USLE model, from which the FER-USLE model was modified. Using the FER-USLE model, strategies applicable to soil loss control in cultivated land with a focus on big events are possible.

(2) Litter performs an important hydrological function in the forest and tea garden. The effect of the litter factor on soil loss depends upon two factors: one is litter areal coverage and the other is the depth of litter. The litter factor differs considerably not only among different types of forest plots but also among the

same types of land use in the Shangshe catchment. Using the litter factor, it is possible to develop strategies for soil loss control in forestland and tea garden with a focus on land with little litter.

(3) Application of the FER-USLE and tests of this model showed that predicted values of *SSD* using FER-USLE agree with observed *SSD* at both the USLE-plot and the sub-catchment scales. The FER-USLE model is applicable for annual soil loss prediction at the catchment scale.

(4) To support policy decisions, the greatest need is for models that can produce robust results using readily available data. The FER-USLE model, as introduced above, is available to many users and policy makers in the Dabie Mountains, China.

References

Kirkby MJ, Imesons AC, Bergkamp G, Cammeraat LH (1996) Scaling up process and models from the field plot to the watershed and regional areas. J Soil Water Conserv 5:391–396

Renard KG, Foster GR, Weesies GA, McCool DK, Yoder DC (1997) Predicting soil erosion by water: a guide to conservation planning with the revised universal soil loss equation (RUSLE). In: Agricultural handbook, vol 703. US Department of Agriculture, Washington, DC

Renschler Chris S, Harbor Jon (2002) Soil erosion assessment tools from point to regional scales – the role of geo-morphologists in land management research and implementation. Geomorphology 47(2–4):189–209

Turner MG (1989) Landscape ecology: the effect of pattern on process. Annu Rev Ecol Syst 20:171–197

Wischmeier WH, Smith DD (1978) Predicting rainfall erosion losses. A guide to conservation planning. In: Agricultural handbook, vol 537. US Department of Agriculture, Washington, DC

Chapter 8
Model of Forest Hydrology Based on Wavelet Analysis

Abstract This investigation proposes a new approach for establishing the relationship between rainfall and runoff, as a function of forest coverage by wavelet-based analysis. First, the Db4 discrete wavelet transform is used to decompose the rainfall and runoff time series to obtain wavelet coefficients at multi-resolution level. The results show that trends of rainfall and runoff were similar. However, runoff did not always follow the trend of rainfall, as it was also influenced by other factors. Second, these wavelet coefficients are applied to model. The results show that potential impacts of forest coverage on hydrological response are of significant importance in Lao Shi-Kan watershed. Runoff decreases as vegetation coverage increases.

8.1 Introduction

Vegetation, especially in the case of forests, plays an important role in regulating runoff, as it reduces dramatically surface water volume, runoff velocity, and peak discharge. Removal of forest coverage causes important changes in the hydrological balance of a watershed, although the magnitude of the response is highly variable and unpredictable. Increased forest coverage, such as the replacing of pasture areas by forest, can result in a reduction of annual flow of up to 40%. It is therefore essential to study the relationship of rainfall, runoff, and forest coverage.

The establishment of a clear rainfall–runoff–forest coverage relationship is difficult due to the large number of variables which affect the process. Pizarro et al. (2006) studied runoff coefficients and their relationship to vegetation coverage and water yield in the Purple River basin, as influenced by land use and the replacement of native forest by plantations. Although there are widely available methods for analyzing and simulating the rainfall–runoff process, most methods and models involve hydrological time series with only the time series data itself. In practice, studying hydrological time series is difficult because hydrology is controlled and influenced by complex factors.

Wavelet analysis has also been applied in the investigation of the rainfall–runoff relationship. The distinct feature of a wavelet is its multi-resolution characteristic, which is becoming an increasingly important tool to process images and signals. Since a wavelet is a localized function in both time and frequency domain, it can be used to represent an abrupt variation or a local function vanishing outside a short time interval adaptively. So, in the field of hydrology, the use of wavelets could be essential in the analysis of rainfall and runoff.

J.C. Zhang et al., *Theory and Practice of Soil Loss Control in Eastern China*,
DOI 10.1007/978-1-4419-9679-4_8, © Springer Science+Business Media, LLC 2011

8.2 Methods

Two methods of analysis are first described: wavelet analysis and runoff modeling.

8.2.1 Wavelet Transform

In recent years, there has been an increasing interest in the use of wavelet analysis in a wide range of fields in science and engineering. Wavelet transform analysis, developed during the last two decades, appears to be a more effective tool than the Fourier transform (FT) in studying non-stationary time series. Of course, this provides an ideal opportunity to examine the process of energy variation in terms of where and when hydrological events occur.

Assuming a continuous time series $f(t), t \in [\infty, -\infty]$, a wavelet function $\psi(t)$ that depends on a non-dimensional time parameter t can be written as

$$\psi(t) = \psi(a,b) = |a|^{1/2}\,\psi\left(\frac{t-b}{a}\right) \tag{8.1}$$

where

t stands for time;
a for the time step in which the window function is iterated; and
b for the wavelet scale.

$\psi(t)$ must have a zero mean and be localized in both time and Fourier space. The continuous wavelet transform (CWT) is given by the convolution of $f(t)$ with a scaled and translated $\psi(t)$:

$$W_{\mathrm{f}}(a,b) = |a|^{-1/2} \int_{t=-\infty}^{\infty} f(t)\,\bar{\psi}\left(\frac{t-b}{a}\right)\mathrm{d}t \tag{8.2}$$

The lower wavelet scales refer to a compressed wavelet, and these allow us to trace the abrupt changes or high-frequency components of a signal. On the other hand, the higher wavelet scales composed of the stretched version of a wavelet and the corresponding coefficients represent slowly progressing occurrences or low-frequency components of the signal.

Calculating the wavelet coefficients at every possible scale is a fair amount of work, and it generates a lot of data. If one chooses scales and positions based on powers of two (dyadic scales and positions), then the analysis will be much more efficient as well as accurate. This transform is called a discrete wavelet transform (DWT). Assuming $a = a_0^j$, $b = kb_0a_0^j$ $a_0 > 0$, and $a_0 \neq 1$, $b_0 \in R$, DWT takes the form of

$$W_{\mathrm{f}}(j,k) = a_0^{-j/2} \int_{-\infty}^{\infty} f(t)\bar{\psi}\left(a_0^{-j}t - kb_0\right)\mathrm{d}t \tag{8.3}$$

The discrete wavelet transform decomposes the input hydrological time series into detailed signals, as an approximation, so the original hydrological time series is expressed as a combination of wavelet coefficients, at various resolution levels.

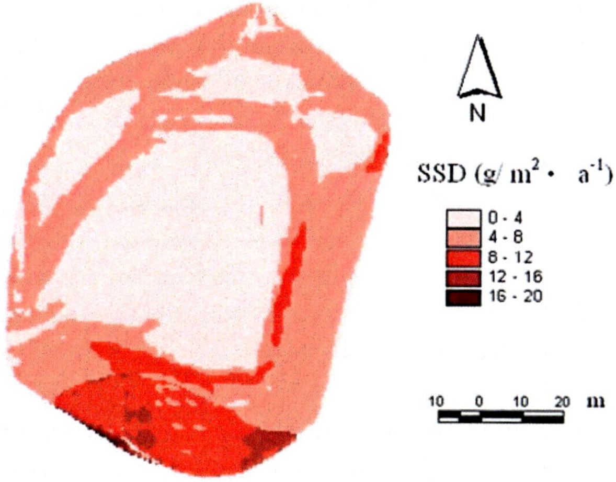

Fig. 5.19 Spatial distribution of predicted annual soil loss in 2007

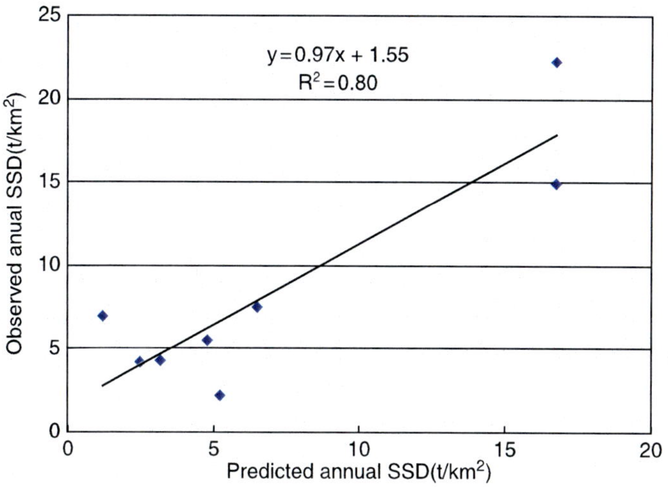

Fig. 5.20 Comparison between predicted annual soil losses and observed losses from years 2000 to 2007

5.7 Conclusions

In this study, field observations of water and sediment discharge were carried out at the micro-plot scale, the USLE-plot scale, and the sub-catchment scale in the sub-catchment of Chinese fir forest and the sub-catchment of pine forest in the Shangshe catchment, Dabie Mountains, China. Through analyses of field observation data in the two sub-catchments from 1999 to 2007, the following conclusions were reached.

Litter provides important hydrological function in forest. In the Chinese fir forest, the micro-plot without litter and grass suffered 71 times more soil loss compared to a micro-plot with litter and grass at the same gradient in 2000.

By integrating a linear regression method with USLE and GIS, a FUSLE model was developed to predict soil loss in forested land. The *R* factor was turned into the R_e factor and the litter was added as a new factor. These innovations, including taking account of litter, made soil loss prediction more accurate in forestland, because inclusion of the litter factor matched the real observations of soil loss in the forest. This approach was also applied to a sub-catchment of pine forest. The test results showed that predicted annual soil loss agreed well with those observed from 2000 to 2007 as indicated by $R^2 = 0.80$. Unlike the USLE and RUSLE models, which need soil loss data over a long period to predict average annual soil loss, the FUSLE model can utilize field observation data in the form of data from micro-plots over a few years. Besides, the micro-plot method is less expensive and simpler to conduct.

However, this model used field observation data only in a sub-catchment. More experiments are needed to test the FUSLE model at the catchment scale. That will be reported in future work.

Acknowledgments The work reported in this paper was undertaken within the framework of a research project funded by the Forestry Department of Anhui Province, China. We appreciate the help of Luo Hongyan, Yang Yan, HuangXiaying, Hu Zhidong, Liu Caihong, Wu Zhongneng, Chu Chengdong, and Hu Bingqi for sediment/sample collection and processing. Collaboration of the Forestry Department of Yuexi prefecture in maintaining the monitoring stations at the Shangshe catchment are gratefully acknowledged.

References

Aerts R (1997) Climate, leaf litter chemistry and leaf litter decomposition in terrestrial ecosystems: a triangular relationship. Oikos 79:439–449

Bacchi OOS, Reichardt K, Sparovek G (2003) Sediment spatial distribution evaluated by three methods and its relation to some soil properties. Soil Till Res 69:117–125

Balci AN (1963) Physical, chemical, and hydrological properties of certain Western Washington forest floor types. Ph.D. Thesis, University of Washington, Washington, DC

Binkley D, Brown TC (1993) Management impacts on water quality of forests and rangelands. General Technical Report RM. USDA Forest Service, Fort Collins

Corner RA, Bassman JH, Moore BC (1996) Monitoring timber harvest impacts on stream sedimentation: in stream vs. upslope methods. West J Appl 11:25–32

Corter C (1938) Surficial runoff and erosion in Java. Tectona 31:613–728

Dickinsonou A, Collins R (1998) Predicting erosion and sediment yield at the catchment scale, Soil erosion at multiple scales. CAB International, Wallingford, pp 317–342

Didham RK (1998) Altered leaf-litter decomposition rates in tropical forest fragments. Oecologia 116:397–406

Griffin ML, Beasley DB, Fletcher JJ, Foster GR (1988) Estimating soil loss on topographically non-uniform field and farm units. J Soil Water Conserv 43:326–331

Herlina H, Ravi P, Widayat CASE (2003) Factors affecting runoff and soil erosion: plot-level soil loss monitoring for assessing sustainability of forest management. Forest Ecol Manage 180:361–374

Kinnell PIA (2001) Slope length factor for applying the USLE-M to erosion in grid cells. Soil Till Res 58:11–17

8.2.2 *Model of Rainfall–Runoff–Forest Coverage*

The forest's hydrological function is mainly manifested in the fractions of precipitation intercepted as opposed to being lost as runoff. The function of forest in intercepting rainfall is first decided by the capacity for holding water by the forest ecosystem, then by the amount of rainfall. The ratio of water intercepted by the forest to total rainfall varies with the amount of rainfall, P. When the rainfall amount is very small, the interception rate, IT/P, may attain 100%, but the interception rate decreases exponentially with increasing gross rainfall amounts. When the rainfall amount surpasses the maximum interception capacity of the forest system, then the excess amount of rainfall drains, or runs off, and is no longer intercepted by the forest.

The rainfall interception rate decreases with the amount of rainfall. This is because the capacity of the forest for intercepting water is diminished at a rate that depends on the rate of rainfall. Liu Chanming published model relating the forest rainfall interception rate and rainfall amount, which assumes that the interception rate is reduced as the rainfall amount increases, and this process obeys the exponential equation:

$$\frac{\mathrm{d}IT\,(P)}{\mathrm{d}P} = a \cdot \exp\left(-K \cdot P\right) \tag{8.4}$$

where

$IT(P)$ is the total intercepted rainfall when the rainfall amount reaches P;
a is the initial rainfall interception rate; and
K stands for the rate at which the rainfall interception rate decreases progressively with the rainfall amount, which has a close correlation with forest coverage.

When the rainfall amount $P \to 0$, this equation becomes

$$\frac{\mathrm{d}IT\,(P)}{\mathrm{d}P} = a \tag{8.5}$$

Because the capacity to intercept rainfall initially has not been filled by the water, the rainfall interception rate approaches 1, or $a = 1$. Then the model can be rewritten as

$$\frac{\mathrm{d}IT\,(P)}{\mathrm{d}P} = \exp\left(-K \cdot P\right) \tag{8.6}$$

At the other extreme, when the rainfall amount increases to a high enough value, or $P \to \infty$, the equation approaches

$$\frac{\mathrm{d}IT\,(P)}{\mathrm{d}P} = 0 \tag{8.7}$$

First, assume the coefficient K is constant; then carry out the integration of the above equation (8.6) to obtain total rainfall interception:

$$\begin{cases} \int_{P=0}^{P=\infty} \mathrm{d}IT\,(P) = \int_{P=0}^{P=\infty} \mathrm{e}^{-KP}\mathrm{d}P \\ IT\,(0) = 0 \end{cases} \tag{8.8}$$

This yields

$$IT\,(P) = \frac{1}{K}\left(1 - \mathrm{e}^{-KP}\right) \tag{8.9}$$

When $P \to \infty$, we obtain the maximum possible rainfall interception loss:

$$ITM = \frac{1}{K} \tag{8.10}$$

According to the water volume balance, runoff (Q) can be defined as

$$\begin{aligned} Q &= P - IT \\ &= P - \frac{1}{K}\left(1 - \mathrm{e}^{-KP}\right) \end{aligned} \tag{8.11}$$

When the forest coverage (α) changes, the potential maximum rainfall interception loss (ITM) can also change. Assuming $ITM = b\alpha + c$:

When $\alpha = 0$, $ITM = c$;
when $\alpha = 1$, $ITM = b + c$.

From (8.10), we have $K = \dfrac{1}{b\alpha + c}$, then

$$Q\,(P, \alpha) = P - (b\alpha + c)\left(1 - \mathrm{e}^{\frac{P}{b\alpha + c}}\right) \tag{8.12}$$

where b and c are parameters of the model.

8.3 Application of the Model of Rainfall–Runoff–Forest Coverage

8.3.1 Study Basin

This work demonstrates the feasibility of applying the wavelet-based method to model the relationship of rainfall, runoff, and forest coverage by selecting the Lao Shi-Kang watershed located at 119°53′–119°14′E and 30°52′–30°23′N, northern Zhenjiang, as the study area. The area of the watershed of Lao Shi-Kang is

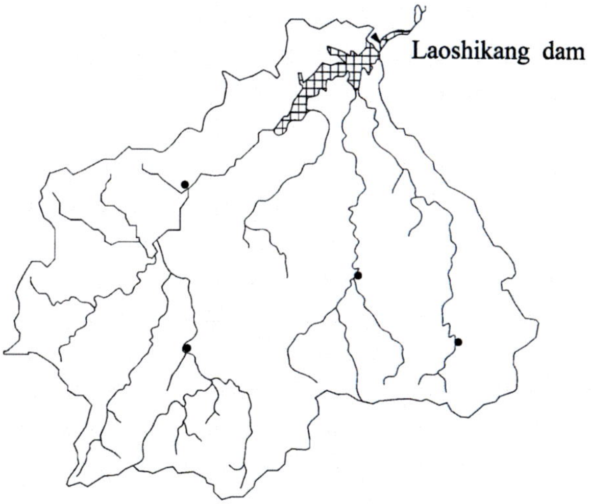

Fig. 8.1 The map of Lao Shi-Kang watershed showing the study area, Anji, Zhejiang

258 km^2, as displayed in Fig. 8.1. The mean annual precipitation in this watershed is 1714.2 mm; the mean annual runoff is 226 million m^3. As for temperatures, the mean annual temperature corresponds to 15°C, with maximum and minimum temperatures of 40.8 and –17.4°C, respectively. There are four rainfall stations and one hydrologic station in the basin (Fig. 8.1).

The time series of yearly mean rainfall and runoff was obtained from the hydrology stations of Anji, Zhejiang. The length of the record is 43 years, from 1962 to 2004. But one of the rainfall stations, Bing-Keng, was missing data from 1996 to 2001. The problem of missing data was solved by interpolation of neighboring points.

8.3.2 Trend Analysis Results of the Wavelet Transform

In this section, we used DW transform coefficients at three decomposition levels. By using discrete wavelet components of time series, we aimed to find which periodicities are mainly responsible for the trend of the series.

The original signal $f(t)$ is decomposed into series of approximations and details following Mallat's algorithm. The process consists of a number of successive filtering steps. The original signal is first decomposed into an approximation (a_1) and accompanying detail (d_1), then the first level approximation (a_1) is decomposed into an approximation (a_2) and accompanying detail (d_2). The decomposition process is then iterated, with each successive approximation being decomposed in turn, so that the original signal is broken down into many lower resolution components.

The time series of the DW transform coefficients represents rainfall and runoff variations. The time series of multi-resolution levels is generated and is presented in Fig. 8.2.

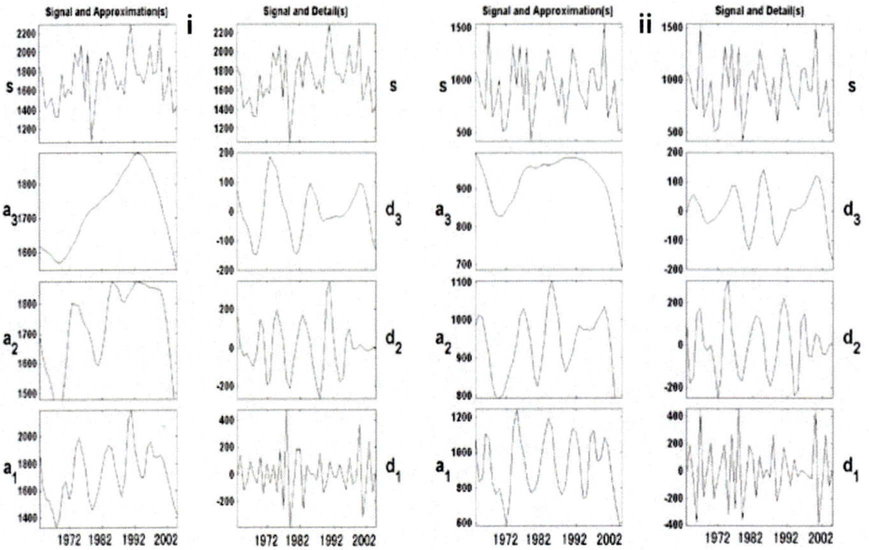

Fig. 8.2 (*Top panel*) Rainfall (i) and runoff (ii) series for the period 1962–2004 for Laoshikang station. (*Other panels*) Time series of first level (a_1, d_1), second level (a_2, d_2), and third level (a_3, d_3)

The top panels in Fig. 8.2 represent the original time series of annual total rainfall (mm) and runoff (mm). The time series of DW coefficient gives the contributions to total annual rainfall and runoff in each year. Time series of first-level mode (a_1, d_1), second-level mode (a_2, d_2), and third-level mode (a_3, d_3) are presented in other panels of Fig. 8.2.

From the first level of approximation (a_1), we can see that the rainfall and the runoff trends are similar on the whole. The second level (a_2) shows a decreasing trend during 1973–1980; however, the runoff increased continuously until 1977. From the third level (a_3), we find that rainfall had a slowly decreasing trend during 1962–1967. Runoff also decreased during this period of time. Rainfall started to increase during the years 1968–1993; the rainfall dropped rapidly after year 1993, the turning point being in 1993. But runoff started to increase after year 1968, then increased gently until year 1977, and then had a declining trend.

8.4 Results of Model

First, we compared the graphs of the original data used in the rainfall–runoff relationship (Fig. 8.3a) and the wavelet coefficients (Fig. 8.3b). It is very difficult to find the actual rainfall–runoff relationships in the original data; so some researchers use a simple linear model to fit it. Applying wavelet coefficients, we can see that rainfall and runoff do not have a unique relationship and thus cannot fit directly into a simple linear model. This is because the forest coverage differed during different

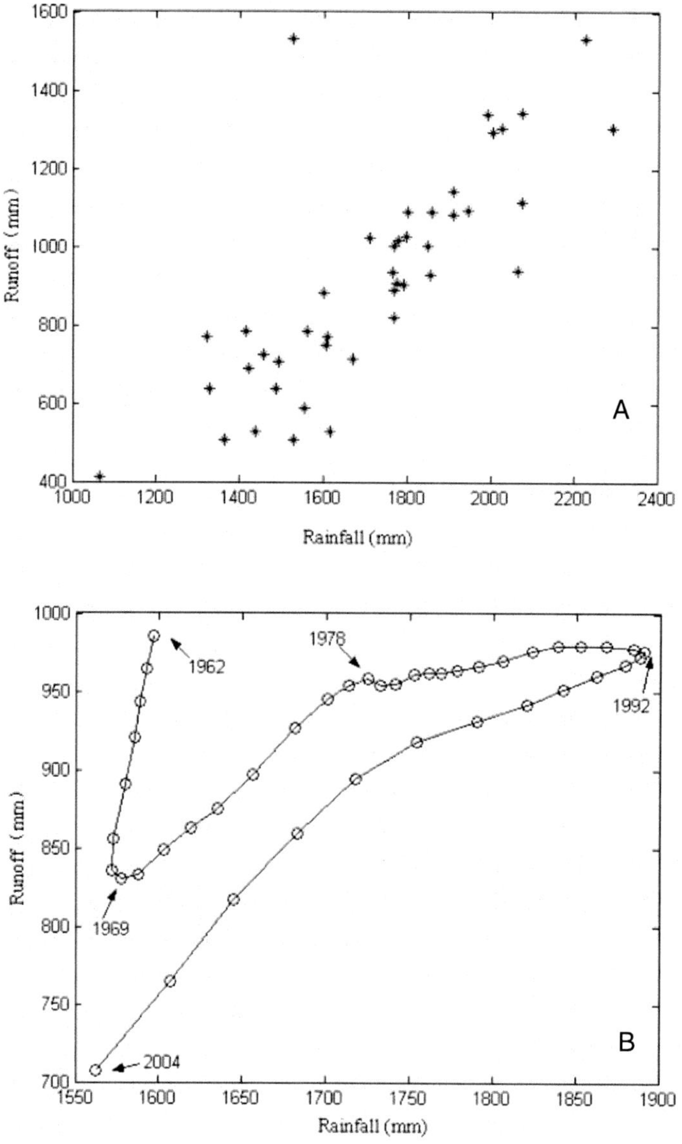

Fig. 8.3 Runoff and the rainfall relational graphs; the original data that was used (**a**) and wavelet coefficients (**b**)

time periods. The runoff of the 1970s was much higher than that of the 1990s under the same rainfall, because forest coverage of the 1990s was higher than that of the 1970s.

Due to the difficulty of forest assessment, it is impossible to get forest coverage each year. Up to now, there have been forest assessments over six time periods in

Table 8.1 Six time periods of assessment of forest coverage in Anji

	First (1953–1957) (%)	Second (1971–1975) (%)	Third (1984–1988) (%)	Fourth (1989–1993) (%)	Fifth (1994–1998) (%)	Sixth (1999–2003) (%)
Whole county	41.00	48.68	52.60	55.90	69.40	71.00
In the basin	64.16	68.77	71.12	73.10	81.20	82.16

China. The recorded data of forest resources in the basin of Lao Shi-Kan are listed in Table 8.1.

The percentages of forest coverage in Table 8.1 represent only the average levels over those periods. In contrast to approximately 40 years of detailed data on rainfall and runoff, the percentages of forest coverage are available for only six time periods. Lacking forest coverage rates for each year, it is certain that the accuracy of the model will decrease, and the difficulty of model parameter estimation will increase. In order to avoid unnecessary errors, we estimate the parameters as follows:

1. For the 40 years of rainfall and runoff, carry out the wavelet transformation separately and obtain the trend component under three different criterion levels;
2. According to the years of the six forest assessments, correspondingly average the trend component of wavelet transformation (Table 8.2);
3. Obtain the corresponding data based on processes 1 and 2, and estimate the model parameters by the least squares method;
4. Apply the data of rainfall and runoff to the model, and obtain the corrected value of the forest coverage of the last 40 years;
5. Average the corresponding years of the corrected value of forest coverage and estimate the model parameters based on this corrected value and the data in Table 8.2;
6. Repeat processes 4 and 5 until the requirements of the model precision are met.

Following the above process, the parameters are estimated using Matlab 6.1, yielding the result $b = 1338.32$, $c = 17.92$, and $R^2 = 0.8856$. Computer programming of the above steps produced simulation results shown in Fig. 8.4.

Table 8.2 Average rainfall and runoff after wavelet conversion

	First (1953–1957)	Second (1971–1975)	Third (1984–1988)	Fourth (1989–1993)	Fifth (1994–1998)	Sixth (1999–2003)
Rainfall (mm)	None	1639.62	1807.92	1876.94	1838.98	1681.66
Runoff (mm)	None	881.80	971.05	976.89	950.15	850.77

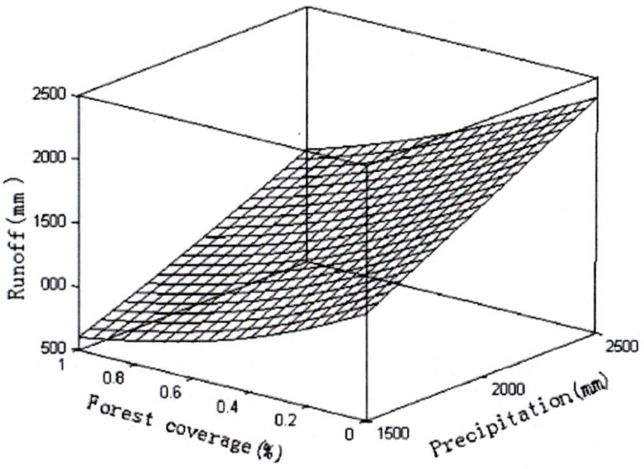

Fig. 8.4 Q–P–α model

Figure 8.5 shows runoff–rainfall relations curves when forest coverage is 0, 50, and 100%, respectively. When forest coverage is zero, or $\alpha = 0$, $Q = P - c\left(1 - \mathrm{e}^{-P/c}\right)$, $K = \frac{1}{ITM} = \frac{1}{b\alpha + c} = \frac{1}{c}$, and $\frac{\mathrm{d}IT(P)}{\mathrm{d}P} = \mathrm{e}^{-P/c}$. Because parameter c is smaller than the normal rainfall, the rate of impounding approaches 0 quickly as P increases so that under 0 forest coverage, the runoff (Q) is approximately equal to rainfall (P). As the forest coverage is increased, the relationship curve of Q–P changes.

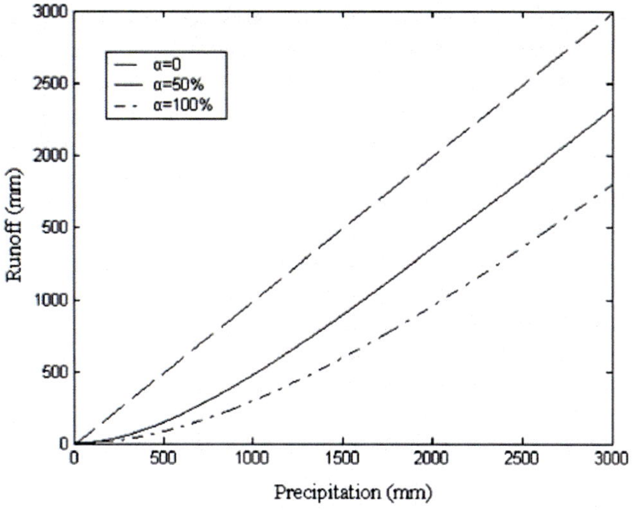

Fig. 8.5 Relation of Q–P, for given values of α

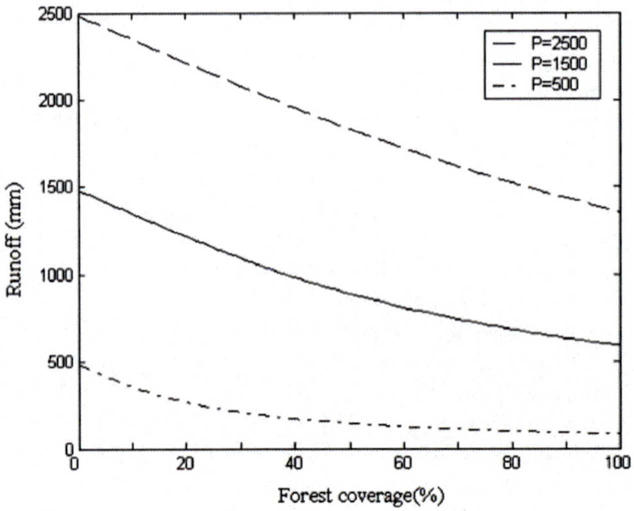

Fig. 8.6 Relation of Q–α, for given values of P

Figure 8.6 gives the runoff–forest coverage relationship curves when rainfall is 2500, 1500, and 500 mm, respectively. When rainfall is 2500 mm (representing periods of abundant water), the runoff is decreased by 1123.7 mm, or 45.27% when the forest coverage is increased from 0 to 100%. When the rainfall is 1500 mm (representing multi-annual means), the runoff is decreased by 889.6 mm, or 60.02% when the forest coverage is increased from 0 to 100%. When the rainfall is 500 mm (representing a dry season), the runoff is decreased by 415.15 mm, or 83.03% when the forest coverage is increased from 0 to 100%.

From Fig. 8.6, we can see that runoff decreases rapidly as soon as forest coverage increases from 0 to 20% when the rainfall is 500 mm. As the forest coverage continues to increase, the rate of decrease decelerates. Runoff barely decreases after forest coverage increases beyond about 40%. It also means that forest coverage of 40% is necessary to impound rainfall for a low flow period. More forest coverage is needed for impounding rainfall when rainfall is at a high level. Unfortunately, it is hard to determine the precise optimal forest coverage. From the figure, we can estimate that about 80% is necessary for impounding rainfall when it is at an average level.

This model can provide the relationship of runoff, rainfall, and forest coverage. Runoff has a downward trend as forest coverage increases. The influence is more prominent under conditions of less rainfall.

Reference

Pizarro R, Benitez A, Farias C (2006) Instantaneous runoff coefficients for Tutuv river basin, Maule Region, Chile. Bosque (Valdivia) 27(2):83–91

Part III
Practices of Soil Erosion Control in Eastern China

As the reform and opening-up policy goes on, the growth of the middle and lower reaches of the Yangtze River as the economic center of China has to be associated with an agenda for sustainable development. Economic development has brought not only enormous material benefits but also negative influences on the environment – the population in this area has reached a level close to its maximum carrying capacity, and the eco-environment has been severely damaged. The environment is assuming critical importance in the 21st century, with degradation of the eco-environment as a potential bottleneck of economic development. To break this bottleneck the relationship between economic development and the environment has to be coordinated. Therefore, how to care about and improve the local eco-environment is an urgent problem, not merely for the middle and lower reaches of Yangtze River but for all of China in the long run.

Chapter 9
Theory of Vegetation Reconstruction for Various Management Types with Different Site Conditions

Abstract According to the management purpose and technical measures, the forest site of the hilly area in middle and lower reaches of Yangtze River was divided into four "management groups": vegetation reconstruction for areas of extreme erosion and degeneration, regeneration and improvement of secondary forests, agroforestry, and commercial forest on good sites. Characteristics of the four groups are described, based on the vegetation reconstruction theory for each of the different site management groups. In particular, for the extreme erosion group, when the external disturbance stops, the degraded community will gradually start its succession toward a dominant community, i.e., progressive succession. Under purposive human intervention, the succession process can be greatly shortened when the following principles are obeyed. (1) Ecological–economic principle: In an eroded red soil area, returning farmland to forest is not only an important means of vegetation restoration and reconstruction but also a national ecological project. We should combine ecology and the economy not only to improve the eco-environment but also to establish a material base for local farmers to shake off poverty, based on the principle of the optimum distribution of resources. (2) The principle of tailoring measures to suit local conditions and to proceed from an assessment of the real situation; (3) the principle of being guided by science and technology, and the ideal of sustainable development; (4) the principle of flexible management and guidance; (5) the principle of making full use of natural forces.

9.1 Site Management Classification

In the Chinese forest site classification system, the hilly area in the middle and lower reaches of the Yangtze River belongs to the northern subtropics. In the forest site basic classification system, it belongs to the agroforestry region of the northern and middle China Plain and the southern hilly region. The soil in this area is mainly red soil and yellow soil. The climate is warm, with the average temperature during the coldest month (January) being 3–8°C and the average temperature during the warmest month (July) being 27–30°C. The frost-free period is 200–300 days, and most broadleaf trees do not defoliate in winter. Evergreen broadleaf forests are distributed throughout this area, and they, together with deciduous broadleaf species, form the evergreen and deciduous broadleaf mixed forests. Typical subtropical zone trees, such as Mao bamboo (*Phyllostachys edulis*), Chinese fir (*Cunninghamia lanceolata*), masson pine (*Pinus massoniana*), oil tea (*Camellia*

J.C. Zhang et al., *Theory and Practice of Soil Loss Control in Eastern China*,
DOI 10.1007/978-1-4419-9679-4_9, © Springer Science+Business Media, LLC 2011

oleifera), tung tree (*Vernicia fordii*), camphorwood (*Cinnamomum camphora*), and nanmu (*Phoebe zhennan*), are widespread and grow well. The forest zones in the mid-subtropical zone have superior natural conditions and cover a large area. They are mainly distributed in the areas of the middle and lower reaches of Yangtze River, such as the southern mountainous and hilly area, which is an important center of timber plantations and an economic forest base in China. It is also the main base of the important southern commercial forest, which is one of the six key centers of forestry engineering in China.

The area in middle and lower reaches of Yangtze River has superior natural conditions. It is rich in agriculture, forestry, husbandry, and industry. The plantation land is important in diverse ways. The forest has great significance for development of the social economy and also provides important materials for industrial construction. Therefore, how to use its production capacity to offer both ecological and economic safeguards for the composite system of nature–society–economy is of great strategic significance.

The regional differences within the eastern part of the southern hilly area consist mainly of landform, lithology, slope position, soil texture, soil thickness, and soil humus thickness; of these, the slope position is most important. If the positions on the slope are divided into upper, middle, lower parts, gully bottom, and valley, then the soil erosion is serious on the upper parts, which have barren and dry conditions, while the middle and lower parts have good water and soil conditions, so the trees grow well. There are also large limestone hills in the study area. Then, based on the bare rock fraction, the slope, and the degree of difficulty of vegetation restoration, the limestone hills can be divided into four site types: extremely difficult to restore, difficult to restore, less difficult to restore, and quite easy to restore.

The site type divisions provide the basis for the forest cultivation and management; therefore, according to the management purpose and technical measures, we divide the forest site of the hilly area in middle and lower reaches of Yangtze River into four management groups:

1. The vegetation reconstruction of extreme erosion and degeneration site management group;
2. The regeneration and improvement of secondary forest management group;
3. The agroforestry management group;
4. The commercial forest on good sites management group.

9.2 Characteristics of the Four Management Groups

9.2.1 The Vegetation Reconstruction of the Extreme Erosion and Degeneration Site Management Group

(1) Vegetation reconstruction in extreme erosion area management subgroup

The rainfall in the southern hilly areas is abundant and unevenly distributed, with rainstorms commonly occurring. Therefore, the steep slopes with serious vegetation destruction, where soil erosion can easily occur, are ecologically fragile areas. Steep slopes exist on the sides of roads, channels, rivers and reservoir watersheds, around residential areas, and in key ecological protection areas. The goal of forest management is mainly to protect vegetation, reconstruct vegetation, and increase vegetation coverage, in order to achieve the objectives of conserving soil and water sources, increasing soil fertility, and conserving landscape functions. This management subgroup should be planted with evergreen deciduous broadleaf forest and coniferous and broadleaf mixed forest with dense crowns and rich deciduous vegetation. The poor site condition forest is a subgroup that should be planted with shrubs such as *Lespedeza* (*Lespedeza bicolor*), *Amorpha* (*Amorpha fruticosa*), and herbs such as Bahia grass (*Paspalum notatum*), and otundus (*Cynodon dactylon*).

(2) Difficult site forest management subgroup
This subgroup includes limestone hills, extreme erosion sites, and tail mining areas. The elution effect on the limestone hills is intense and, as a result the soil is thin, the fraction of bare rock is high. The soil can be neutral, alkaline, or may have appreciable acidity, but the soil calcium content is high, and it may require special forest management measures to enable trees to be planted.

Extreme erosion means there is serious soil degradation, mainly manifested in a heavy loss of surface soil layer. The original vegetation no longer exists, the soil erosion is severe, and there is nearly no humus layer. The tail mining area is an exceptional site, with tailings piled up without the basic soil characteristics. Reconstruction of vegetation requires special measures.

9.2.2 The Regeneration and Improvement of Secondary Forest Management Group

The hilly area in the middle and lower reaches of the Yangtze River has a broad distribution of secondary forests. They are rich in tree species resources and have high vegetation coverage. However, many of the existing forests have dominant trees that are small and have low productivity, inefficient management, and low utilization.

Applying regeneration and transformation measures to the fine saplings such as sweet gum (*Liquidambar formosana*) and red *Schima superba* (*Schima wallichii*), camphorwood and so on can create a good environment, high productivity, and achieve a good cultivation effect.

9.2.3 The Agroforestry Management Group

In the hilly agricultural area, there often exist conflicts regarding allocation of land among different purposes: agriculture, forestry, husbandry, and management. The intensive agroforestry and stereoscopic management (utilizing multiple vertical layers on land to achieve more than one objective) is an effective way to resolve conflicts and to coordinate development of the economy and the environment. It is also a powerful measure to coordinate the relationship between short-term benefit and long-term benefit.

9.2.4 Good Site Commercial Forest Management Group

Forests in the "good site commercial forest management group" are located in low mountains and hills. Their site conditions are superior and are often characterized by soil thickness and humus layers that are medium or thick, and soil chemical properties that are good. The original forests, which are usually distributed in valleys, downhill areas, or gentle slopes, are luxuriant.

According to the needs of economic development, the aims of forestry management in each area, and the habits and experience of the local people, this type of forest may be divided into management subgroups of timber forest, industrial raw material forest, and economic forest.

(1) Management subgroup of timber forest and industrial raw material forest
Low mountains and hills are usually remote areas far away from towns. Because of this, there are large areas of forest that can provide timber and industrial raw material, and are especially suitable for the development of fast-growing, high-yield forest, with large diameter-class timber or rare tree species timber. Their main tree species are Chinese fir and Mao bamboo, and broadleaf tree species, such as Fagaceae, Theaceae, and Magnoliaceae. In such sites, trees can achieve a higher growth rate and farmers can obtain comprehensive benefits from the forest.

(2) Subgroup of economic forest management
In the hilly areas near settlements, residents have planted mainly economic forest types, such as oil tea, shoot-producing bamboo, tung tree, *Ginkgo* (*Ginkgo biloba*), chestnut (*Castanea mollissima*), red bayberry (*Myrica rubra*), pecan (*Carya cathayensis*), and *Torreya* (*Torreya grandis*). These are typical subtropical economic trees.

9.3 Vegetation Reconstruction Theory of Different Site Type Management Groups

9.3.1 Management Group of Extremely Eroded and Degenerate Inferior Lands

(1) Principles

Barren hills, a typical subtropical degraded ecosystem, result from the retrogressive succession of forest communities influenced by human activities or natural disturbances. It is a manifestation of environment deterioration. The large degraded ecosystem of barren hills has seriously restricted the functioning of ecological service systems, resulting in ecological catastrophes that even hamper the development of the local economy.

When the external disturbance stops, the degraded community will gradually start its succession toward a dominant community, i.e., progressive succession. Under purposive human intervention, the succession process can be greatly shortened if the following principles are obeyed:

(i) Ecological–economic principle
In an eroded red soil area, returning farmland to forest is not only an important means of vegetation restoration and reconstruction but also a national ecological project. We should combine ecology and the economy not only to improve the eco-environment but also to establish a material base for local farmers to shake off poverty, based on the principle of the optimum distribution of resources.
During the implementation of the project, it should be realized that the eco-environment is also a kind of resource and wealth. Some of its functions are obvious and direct and some are only potential or indirect. Besides, it is also the foundation and carrier of other types of materials and spiritual wealth, as well as being the basis for the sustainable development of economy and society.
(ii) The principle of tailoring measures to suit local conditions and to proceed from an assessment of the real situation:

  (a) The principle of planting vegetation appropriate to the environment
  It is the key premise for vegetation restoration and reconstruction. The design and implementation of the ecological project must be based on the specific conditions of the area.
  (b) The principle of respecting the willingness of local farmers
  During the implementation of vegetation restoration and reconstruction, we should fully respect the willingness of local farmers and make sure their short-term and long-term benefits can be fully guaranteed.

(iii) The principle of making full use of natural forces
Usually, the areas for vegetation restoration and reconstruction are sloping farmland with extremely difficult site characteristics. The ability to improve the site conditions by artificial measures is very limited. These objective conditions determine the means of vegetation restoration for which natural forces should be utilized as much as possible, while fully considering the natural attributes of agricultural and forestry production.
In other words, regional environment, site conditions, and production capacity should be comprehensively considered in order to implement vegetation restoration and reconstruction only in the areas suitable for forest growth, and to make full use of modern forest cultivation – these exemplify the rational use of natural forces.
(iv) The principle of being guided by science and technology, and the ideal of sustainable development

As the primary forces of progress, science and technology are also essential for the ecological projects of vegetation restoration. For example, scientific and technical judgment can be shown in the selection of tree species, sturdy seedlings, afforestation technology, and even business strategy formulation. Sustainable development, which should be reflected and implemented in concrete measures and technology, not only is a strategic target but also manifests itself in each specific working process.

(v) The principle of flexible management and guidance
Because of the greatly varied conditions in the hilly area in middle and lower reaches of Yangtze River, flexible management and guidance should be applied based on different social and economic conditions, management direction, management custom, and the site types.

(2) Guiding ideology

According to ecosystem characteristics of different types of degraded barren hills, the restoration and reconstruction objectives need to be determined appropriately. Then the forest layout can be arranged and tree species can be selected based on the main objectives (soil erosion control, rapid vegetation restoration, promoting relict forest succession progress, and improving the woodland environment) and principles (growing trees in suitable places, paying attention to the balance of long, medium and short-term benefits, and sustainable development). Trees, shrubs, and grasses should be grown together; coniferous and broadleaf trees should be combined to attain a mixed forest, with agriculture and animal husbandry. Equal attention should be paid to both ecological and economic benefits, and scientific research should be closely related to production by making use of advanced cultivation techniques and the latest research results, establishing many experimental combinations, and selecting the best treatment and combining engineering measures of soil and water conservation.

(3) Aims and measures

Adopt corresponding measures to restore and reconstruct vegetation according to different types of barren hills:

(a) Severely eroded barren hills
The barren hills that were studied are located in Guifeng and Gan counties of eastern and southern Jiangxi Province, and were intensely eroded and destroyed. The main objectives of barren hill governance were vegetation restoration and soil erosion control. Trees tolerant of drought and barren conditions, such as masson pine and loblolly pine (*Pinus taeda*), and soil-stabilizing grasses, such as *Roegneria kamoji* (*Elymus kamoji*), crabgrass (*Digitaria sanguinalis*), and Dallis grass (*Bothriochloa intermedia*), that is, a rational selection of trees, shrubs, and grasses, were chosen to build a coniferous and broadleaf mixed forest. By making use of advanced silviculture techniques (ABT, rooting powder, container seedling, superior provenance, etc.) and measures such as fertilization and afforestation at high densities, a high survival rate was ensured.

For steep study sites, water and soil conservation projects, such as site preparation of horizontal ditches and excavation of "fish scale"-shaped pits, were carried out.

(b) Barren hills of limestone bare rock

The barren hills of limestone bare rock that were studied are in Xiushui prefecture in western Jiangxi Province. In view of the thin but relatively fertile soil, which is nearly neutral in pH, there are three major objectives: (1) forest restoration; (2) the combining of governance, development, and utilization; and (3) the establishment of a forest–agro-pastoral composite business model. Tree species, such as Chinese arborvitae (*Platycladus orientalis*), Sichuan alders (*Alnus formosana*), *Chinese cedar* (*Toona sinensis*), *Eucommia* bark (*Eucommia ulmoides*), and *Wheel wingnut* (*Cyclocarya paliurus*), which not only are adapted to limestone but also have economic value for stereoscopic combination, were selected. Crops and forage grass like grain amaranth (*Amaranthus hypochondriacus*) were intercropped to develop a cultivation industry.

(c) Gravel thatch barren hills

These barren hills that were studied are located in Yujiang and Taihe prefectures of eastern and central Jiangxi Province. The main objectives were to improve land capability, to restore vegetation, to establish a fuel forest base, and to pay equal attention to long-, medium-, and short-term benefits. Masson pines were selected as fuel forest management species, while multipurpose forest was established and agroforestry management in the gully bottoms was carried out.

(d) Dense brush barren hills

These barren hills that were studied are located in Jingdezhen City, Pingxiang City, and Yiyang City in northern, western, and eastern Jiangxi Province. The main goals of rebuilding these barren hills were to artificially promote the progress of succession, to rapidly restore forest communities, and to develop forests for timber and industrial raw material. For forests that had a certain amount of fine broadleaf trees growing among shrubs, the finest broadleaf tree species were selected, slash pines (*Pinus elliottii*) and pond pines (*Pinus densiflora*) were planted, and management was subsequently carried out. For Pingxiang and Yiyang cities, commercial tree species were replanted, which included mainly spruce, slash pine, masson pine, and loblolly pine; the existing broadleaf trees were protected, raised, or interplanted tung tree, *Eucommia* bark, and other short-term yield commercial trees. Interplanting experiments of Guangdong purple bead (*Callicarpa kwangtungensis*) for shade-tolerant medicinal materials were carried out among young Chinese fir plantation at an age of 2 years.

(e) Barren hills with thin shrubs

Barren hills with thin shrubs were studied near towns and villages in Le'an and Jinxian prefectures in central Jiangxi Province. These areas have flat terrain, thick and relatively fertile soil, and are subject to a high intensity of plowing and wood chopping. The main objectives of ecological restoration were to increase the vegetation cover, to establish the intensive agroforestry ecosystem, and to develop a suburban forestry relevant to the towns.

On this type of land, not only a mixed forest consisting of slash pine, masson pine, red *S. superba*, sweet gum, south wild jujube (*Choerospondias axillaris*), and Sichuan alders but also a fire-proof wood forest of pitch pine (*Pinus rigida*) was constructed. Furthermore, *Cedrela sinensis* forest for use as a vegetable, a fruit forest of chestnut (*C. mollissima*), orange, and Nai plum, and a medicinal forest of *Eucommia* bark, *Magnolia officinalis* (*Houpoea officinalis*), Yellow gardenia (*Gardenia stenophylla*), oil tea, and tung tree were also successfully planted. In addition, crops and vegetables were interplanted among young trees to gain short-term profits by farmers, which in turn improved the soil fertility and facilitated the growth of young trees.

9.3.2 Basic Theory of Vegetation Restoration and Reconstruction in Limestone Hills

There are many limestone hills in the hilly area in the middle and lower reaches of the Yangtze River, among which is an area in Jiangxi Province of 204,500 ha, 1.709% of the entire province territory. There are 116,900 ha in Jiujiang City, 45.9% of the total area of the barren limestone hills in Jiangxi (Photo 9.1). Afforestation and revegetation in these kinds of hills is difficult because of the exposed rocks. Therefore, 166.67 ha of barren limestone hills in Xiushui County of Jiujiang City was selected as a research object. In view of the characteristics of barren limestone hills and the special difficulties in afforestation work, stereoscopic management and forest–agriculture–herd patterns were adopted in view of eco-economic goals.

Photo 9.1 Limestone hills in Jiujiang City in Jiangxi Province

(1) Guiding ideology

By using ecological and economic theories and methods, and making full use of forestland and other environmental resources (light, heat, water, air, and fertilizer) within a certain period of time and limited space, forestry, agriculture, and animal husbandry were combined at multiple levels to establish a multi-component, multi-level, multi-sequence, and multi-function eco-economic system that uses the forest as its main base.

(2) Principles

According to the natural features and local economic development status in limestone hills, the development and management should be based mainly on the following principles:

(a) The principle of combining ecological, economic, and social benefits
Because of poor natural conditions in barren limestone hills, priorities should be given to afforestation, vegetation restoration, environmental improvement, and ecological equilibrium. However, at the same time, enough attention should also be paid to the local economic development and benefits to the populace, i.e., both ecological functions and economic benefits should be achieved.
(b) The principle of multi-dimensional management and multi-layer structure
Because of different soil depths and great varieties in natural conditions from the catchments to the foot of the hills, in order to fully increase the production of forests, multi-dimensional development should be considered by planting trees according to specific time and site. For example, the limestone soil has high fertility. When the small trees were newly planted, the inter-spaces between them were large enough to allow the creation of agroforestry systems in these places, which in turn facilitated the growth of trees.
(c) Principle of multistage utilization of material circulation
The combined system of forestry, agriculture, and animal husbandry transforms the organic matter originally created by primary production into products with higher economic value, achieving its functions of enriching, promoting, and maintaining the structure of the food chain. Finally, it achieves the multi-utilization of materials and effectively improves the economic value of the system by artificially increasing the various components of the production chain.

9.3.3 Agroforestry System Group

(1) Tea–tree intercropping system

(i) Ecological basis of the tea–tree intercropping system
Coordinating the relationships among plant populations in an ecosystem to a large extent depends upon the spatial relationships among the individual plants, and the life histories and ecosystem functions of each species in the system:

(a) Spatial structure basis
In pure tea gardens, spatial structures both above and below ground surface are simple single layers. However, in the compound management of tea

gardens combined with trees, their spatial structures are horizontally interspersed in an efficient way and vertically layered above the ground surface formed by a combination of high and light-loving arboreal species and tea, and short and shade-tolerant trees.

(b) Temporal structure basis
In the ecosystem, temporal successional sequences of different species groups often occur in several ways: temporal overlap and partial temporal overlap. Overlap of different species groups in time results in competition for environmental resources. Partial overlap and intercross may be used in providing coordination, complementarity, and adaptability between these different groups in their demands on environmental resources.

(c) Basis in functional roles
In ecosystems, different populations' functional roles are often referred to as their either direct or indirect functions. In the tea–tree intercropping system, the direct function is to produce lumber, tea, and other forest products, which have many varieties and are extensively used to meet various needs. The indirect functions stem from such relationships as the shade provided by the arboreal species, the interweaving of root systems, and the increasing of the quantity of leaf litter. There are great differences in the micrometeorological factors and soil conditions between tea–tree intercropping gardens and pure tea gardens.
In summary, species groups in interplanted tea gardens supply and complement each other in space. When there is partial spatial overlap, competition is not fierce. In terms of functions, this kind of system not only enriches tea garden ecosystems but also benefits the growth of the trees. In a word, there are mutualistic niche relationships among species groups in the tea–tree intercropping system.

(ii) Influence of tea–tree intercropping on tea production

(a) Influence on tea yield
A five-year on-site experiment shows that, in a slash pine–tea intercropping garden, the average yield is 6.5% higher than that of a pure tea garden. Degree of shading, or "shading rate," which in a pure tea garden is zero, has great influence on tea yield. As determined from multiple comparisons, the relationship between tea yield and shading exhibits the following pattern: systems with shading rate between 30 and 40% have the highest yield; those with shading rate between 20 and 25% are the second; the next are those with a shading rate of 0% (pure tea garden); and the least yield belongs to those with shading rates between 45 and 50%. This shows that it is good for the pine–tea intercropping system to have a shading rate between 30 and 40%, while higher shading rates (above 45%) may have a negative effect on tea trees' photosynthetic process. Therefore, attention should be paid to the density of pine trees in pine–tea intercropping systems.

Table 9.1 Comparison of tea quality between two types of tea gardens

Types of tea gardens	Amino acid (g kg^{-1})	Caffeine (g kg^{-1})	Soluble sugar (g kg^{-1})	Water extract (g kg^{-1})	Flavonoids (g kg^{-1})
Intercropping tea garden	32.8	36.4	14.7	58.1	6.94
Tea garden without intercropping	27.4	34.1	13.1	49.9	11.85

(b) Influence on tea quality

A good tea garden ecological environment influences not only the yield of tea but also tea's biochemical substances and their proportions, and, to a large extent, the appearance of the tea. Biochemical analysis (Table 9.1) revealed that tea plants in a tea–tree intercropping garden have more amino acids, caffeine, soluble sugars, etc. than do those in a pure tea garden, which shows that the ecological environment in a compound management tea garden can improve the biochemical quality of tea.

(2) Agroforestry systems

Agroforestry systems are essential to solving problems on the way to sustainable development of the forest regions and for promoting economic development. Such systems include the following: planting fruit or nut trees, or introducing forage grass, increasing or maintaining the yield of crops indirectly by improving soil fertility and agricultural sustainability; benefiting the output of crops by improving soil structure and increasing soil fertility; reducing soil and water loss; promoting the formation of micrometeorology suitable for both agriculture and forestry; providing various agricultural products and services; promoting the development of stockbreeding by means of intensified forage grass production and livestock raising; and making full use of resources to plant Chinese traditional medicines, flowers, edible fungi, and other kinds of forest by-products. Generally speaking, agroforestry management can help to increase and keep the organic matters and the nutrient contents in the soil and promote recycling, and improve physical and chemical conditions of the soil. In the following, the cultivation of grain amaranth will be used as an example to describe this system in detail.

Grain amaranth is a kind of high-quality protein feed resource. With rich nutrition, strong resistance, quick growth, high yield, strong germination ability, good nutritional quality, and extensive use, it is known as the most promising crop in the 21st century. Since being introduced from the American Rodale Organic Gardening and Modern Agriculture Research Center in 1982, it has proven to have good compatibility in the red soil of hillslope land in Jiangxi Province and also to have wide application in economic forests, orchards, and other agroforestry systems.

The good biological and ecological characteristics of grain amaranth have established its important ecological niche in agroforestry. At present, management design methods have applied it in eco-agriculture. Eco-forestry and agroforestry mainly follow the principles of allocation of time–space and nutrition relations by which is meant the interplanting based on the arrangement of spatial position and temporal sequencing, so as to fully utilize climate and soil resources, and the manipulation of the trophic structure, to utilize both the primary production and secondary production processes of the food chain. In this way, substances and energy are fully utilized and even waste material is turned into resources. Experience indicates that many compound management patterns show satisfactory ecological rationality in terms of guiding principles. However, the operation of many of these patterns is still in the phase of testing or experimentation on a small scale and thus do not yet fully demonstrate the vitality of eco-agroforestry systems. This lack of demonstrable effect has been caused, to a large extent, by the neglect of the demands of market disciplines, namely economic rationality, during the design phase. For example, in the southern region of Jiangxi Province, a pig–biogas–orchard project was built to produce more pork by raising pigs and simultaneously by using the manure produced by pigs to fertilize the orchard soil or produce biogas, the fluid of which is also a good fertilizer. In this way, meat and fruit could be produced continuously, and the energy problem in the countryside could be solved. Many counties and cities developed their fruit industry according to this pattern, planting fruit trees and building pigsties. However, the results achieved have rarely been satisfactory. Using commercial feed to raise pigs has had to be given up because the cost exceeded the price of pigs, causing the entire production chain to shatter or work at low efficiency. Many other similar planting–cultivation chains have also suffered from high costs.

The grain amaranth–forestry agroforestry system can improve this situation, as the introduction of grain amaranth makes the flow of substances and energy in the system more efficient. It expands and optimizes the biological food chain. Multiple uses of matter are created in the system: light and nutrients in the forest (orchard) are fully utilized; large quantities of high-quality feed are provided for animal husbandry; tending is replaced by tillage; soil is improved and protected; micrometeorological effects are improved to promote the growth of forest (fruit trees); in addition, waste materials may also be used to create both energy and high-quality fertilizer. A particularly important thing is that grain amaranth can be a substitute for concentrated feeds and help to reduce the cost of cultivation. Meanwhile, it can be easily grown at a low cost, which increases the organic manure needed by the forest for fruit production, and it effectively promotes the development of the breeding industry, and thus makes the forest (fruit) industry more prosperous. The system is also benefited by maximizing biomass production, which brings returns back to the forest by way of livestock production, and thus expands the use of the food chain. More products bring more money, which is again invested into agriculture, forestry, and fruit planting in order to make greater profits.

In fact, grain amaranth functions as the link between the forest, agriculture, and animal husbandry. It fills an empty ecological niche in the forest and forms the connection between the basic nutrition level and secondary production in the material circulation of the forest–agriculture–herd ecosystem, and consequently makes up for various insufficiencies in the development of the forest–fruit industry at present. The practice proves that, as a kind of high-quality feed source for livestock and good component in the forest intercropping pattern, grain amaranth may make full use of its favorable characteristics, such as a short cycle and ability to be used quickly, and solve the problem of the long cycle and slow benefits from forest management. In addition, it may help to maintain other species in the cultivation chain in the agriculture–forestry–husbandry system.

(3) Forest–Chinese herb compound system

Forest–Chinese herb management is an efficient way to improve economic performance of the forest. Medicinal plants were interplanted in a young Chinese fir forest, which has different canopy densities during growth, resulting in different micrometeorological environments. Many medicinal species were selected to make full use of space and to increase short-term profits, which can solve problems of forestry production, such as long cycles, slow development toward usable production, and high investments in stand tending. In the next section, Guangdong purple bead and grass coral (*Sarcandra glabra*) are introduced as examples to describe forest–medicine systems.

(i) Influencing factors on the yield of Guangdong purple bead in intercropping forests

Guangdong purple bead, belonging to *Callicarpa* in the Verbenaceae family, is distributed naturally in the elevation range between 100 and 850 m. It is a kind of perennial shallow-root undershrub and a traditional Chinese medicine with high quality. The whole plant can be used as medicine and has the functions of antipyretics, detoxication, blood cooling, and hematischesis, and has anti-inflammatory, anti-cancer, etc. properties. In consideration of the biological characteristics of Guangdong purple bead and of local management and financial situations, together with the requirements for planting on barren hills, research is being carried on the interplanting plot of young Chinese fir and Guangdong purple bead.

Guangdong purple bead was planted at five different density levels in a 3-year-old forest of young Chinese fir (Table 9.2). Results show that yields varied remarkably and those planted at 30 cm × 30 cm area per plant had the highest yield. Altitudes greatly influence the yield of Guangdong purple bead (Table 9.3). The highest yield occurs at an altitude between 350 and 410 m

Table 9.2 Annual mass yield of Guangdong purple bead at different levels of density (area per plant)

Test area	1	2	3	4	5
Density (cm^2)	20×20	30×30	30×50	50×50	70×70
Mass yield ($t\ ha^{-1}$)	7.51	7.82	4.93	4.35	2.06

Table 9.3 Annual mass yield of Guangdong purple bead at different altitudes

Test area	1	2	3
Altitude (m)	410	350	250
Mass yield ($t\ ha^{-1}$)	8.44	8.00	3.80

Table 9.4 Annual mass yield of Guangdong purple bead at five levels of canopy density

Test area	1	2	3	4	5
Canopy density under Chinese fir forest	0.7	0.6	0.5	0.4	0.3
Mass yield (t ha^{-1})	6.06	6.80	5.69	4.42	2.92

and it decreases with decreases in altitude, which is in accord with the ecological characteristics of Guangdong purple bead, which favors a shady and humid environment.

Since 1996, remarkable economic efficiency has been achieved by interplanting Guangdong purple bead in new young Chinese fir forests for barren hill afforestation. Upper layer canopy density is one of the leading factors that affect the growth of Guangdong purple bead. Annual mass yields of Guangdong purple bead at different levels of canopy density are shown in Table 9.4. Among the five levels of canopy densities, densities between 0.5 and 0.7 of full canopy closure resulted in the most satisfactory yields, and the most suitable canopy density was 0.6.

(ii) Factors influencing the growth of grass coral in intercropping forests

Belonging to *Sarcandra* in Chloranthaceae family, grass coral is a kind of broad spectrum traditional Chinese medicinal material whose artificial cultivation has aroused people's interest. Based on grass coral growth characteristics, multi-use of forest space can be adopted to increase the productivity of the forest. A cultivation experiment was carried out by interplanting grass coral in a Chinese fir forest in the Lishan forest center of Xingan, Jiangxi Province. It proves that interplanting grass coral in forests can not only increase the utilization rates of the forest space but also maintain a necessary natural environment for the growth of grass coral and consequently improve the quality of the medicinal material.

(a) Influence of canopy density on the yield of grass coral

Biologically speaking, grass coral should be grown at a proper canopy density, neither too high nor too low. Experiments were performed on random samples under forest canopy with five different levels of cover density: 0.5, 0.6, 0.7, 0.8, and 0.9. The results, shown in Fig. 9.1, demonstrate that when canopy density was 0.7, its speed of growth was the fastest. According to variance analysis, when the canopy density was between 0.6 and 0.9, the differences among treatments of the heights and of the average diameter growth rates at ground level of grass coral were not great.

(b) Influence of slope position on the yield of grass coral

Because interplanting is done along the contour, in order to check the influence of slope positions on the growth of plant, 10 random samples were picked from the upslope and another 10 from the mid-slope regions. The average height and the average ground diameter of the plants were

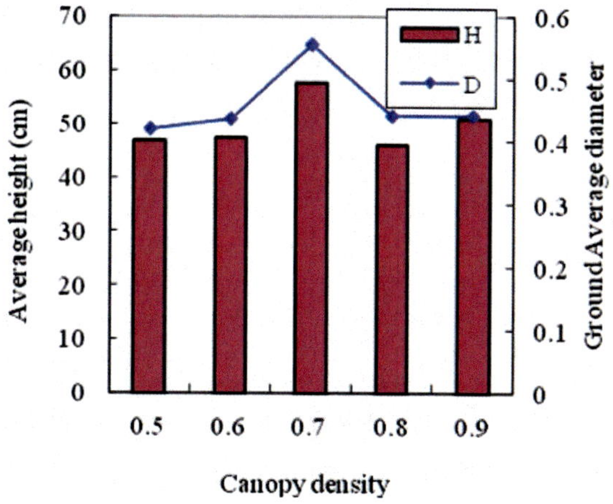

Fig. 9.1 Growth index of *grass coral* at five levels of canopy density

measured. The result of variance analysis indicates that growth varies greatly at different slope positions and that the mid-slope is more suitable for the growth of grass coral than is upslope.

9.3.4 Good Condition Commodity Forest Management Group

In order to solve the problem of ecological development and timber production, and to thoroughly boost the development of the timber forest base construction, planting of timber forests or forests for industrial material and economic forest management can be done separately in stands of good condition and low soil erosion.

(1) Density regulation of timber forests and industrial material forest

The density of planting is the most important aspect of forest management directly influencing the results of forestation: the stand structure, forest growth, rate of use forestland, and forestation cost.

(a) Influence of planting density on the crown growth of trees

Planting density immediately influences the crown and the crown's growth, including crown diameter, crown length, and canopy density or overlap factor. In the 3rd to 5th years after interplanting at different densities, there was no obvious variance of the crown factor. There was, however, a remarkable difference of the crown factor as a function of canopy density in other years under certain conditions. The crown of slash pine is narrower than that of loblolly pine. Its branch extroversion growth is inferior to that of the loblolly pine, its canopy closing is slow, and its environmental pressure is greater than that of loblolly pine. Under the same conditions, the overlap of loblolly pine is larger than that of the slash pine. At 6–7 years of age, in most sample sites where the density of slash pine is 2000/ha and the density of loblolly pine is 1667/ha, the degree of overlap has surpassed 1.0, indicating that the stand is in a competitive stage of development.

(b) Relationship between planting density and growth in height

There is not a remarkable difference in the average height between different planting densities for slash pine and loblolly pine forest in the 6th–7th year after planting. The slash pine's height reaches a maximum when planted at (2.0 m × 2.0 m)– (2.0 m × 3.0 m). In the southern Anhui area, the average tree height of slash pine basically increases as density increases, while in the northern Anhui area, it is the opposite, that is, the height decreases when the density increases, which probably depends on when the competition between trees begins to occur and on the site differences.

(c) Relationship between planting density and DBH increment

For 6–7-year-old loblolly pines, the diameter at breast height (DH) is greatest when the planting density is (2.0 m × 3.0 m)–(2.0 m × 2.5 m). For high density, there is competition between individual trees, which has a positive effect on growth, while at low density, the stands were not dense and there was little competition, so the DBH was not as great as that at (2.0 m × 3.0 m)–(2.0 m × 2.5 m). For a slash pine forest of 6–7 years, the DBH increased as the density decreased. Planting at (3.0 m × 3.0 m)–(2.0 m × 3.0 m) would achieve the highest DBH. Comparing two trees, under the same density, the slash pine's crown and the overlap extent are bigger than those of the loblolly pine. High-density forest entered the competitive stage earlier.

9.4 Closing Hillsides with Secondary Forest to Culture Forest with Least Human Interference and Regeneration Management Group

9.4.1 Theory of Closing the Hillside and Regenerating the Secondary Forest

The hilly region in middle and lower reaches of Yangtze River has good natural conditions and is rich in natural forests. But because of the long period of cutting for small diameter logs, the area of sparse wood and shrubs has increased year by year. Although the hills are all green, they cannot grow forests and have low economic benefit. For this kind of hillside, because the forestation cost is high, closing the hillside against further use for a time and regeneration of the secondary forest method should be used.

1. Theoretical basis of multi-layered structure in closing hillside and forestation

   The experiment place is in the secondary broadleaf forest that has undergone multiple regeneration after several severe destructive events. Vegetation types in this kind of forest are diverse. Its original vegetation was evergreen or deciduous broadleaf forest. Because of the cutting and artificial destruction over a long period of time, its reforestation is difficult. Because of the introduction of masson pine, Chinese fir, sweet gum, and red *S. superba*, these forests need a long period of being closed before they can grow into mature evergreen broadleaf forests. So it is necessary to use tending, replanting, or other eco-economic measures to shorten this process so as to save some of its natural forest characteristics to form a multi-layer crown, to maintain its original condition as far as possible, and to strengthen its stability in fitting into the environment. During the closing of hillsides and replanting of forests, because of the diversity of species and their complex structures, the growth rates and biomasses are also large. At the same time, the populations have a strong compatibility with the external environment.
2. Tree selection

   This study selected the loblolly pine and pond pine as the trees for replanting. At the same time, sweet gum, red *S. superba*, *Alniphyllum fortunei* (*Alniphyllum fortunei*), and the Fagaceae family trees were chosen as trees to be conserved among those that already exist, considering their biological characteristics, compatibility, the overall benefit, and cultivation goal. The goal is to develop forest community types according to the rules of natural succession and to satisfy human economic goals, especially to meet the eco-economic demands of highly effective forestry and forestry-sustained development.

   Both the loblolly pine and the pond pine are fine coniferous species which were introduced to China from overseas. They have characteristics of broad compatibility and fast growth and high yield. The trees planted in a large area of Jingdezhen City in Jiangxi Province showed adaptation to the local climate and had bigger growth compared to the masson pine in the hilly area. The characteristics of pond pine are strong sprouting, straightness, and a full, round trunk, which are unique among all kinds of pines. In Jingdezhen, a city lacking in fuelwood and coal, it can be used both as timber and fuelwood, and it is also good for the conservation of water and soil. Other reserved evergreen trees also provide timber or fuelwood forest of high economic value, and the deciduous broadleaf trees, in particular, have positive effects on the fertility of the forest soil. Replanting loblolly pine and pond pine in the secondary broad-leaf young bushes forms a multi-layer forest structure, which fully utilizes the energy of light and soil fertility, promotes forest growth, enhances the ability to resist disease, improves the quality and yield of the forest, forms a stable forest community, and transforms the low productivity timber woods into a highly effective forest of timber, fuelwood, and protection.
3. Technology of closing hillside and replanting

   In the barren hills there are mainly broadleaf bushes that have germinated for several generations after a long period of cutting. They have low density and are distributed spatially homogeneously, without any big trees. The following technologies are used to create a forest very rapidly, using the process of closing hillsides and replanting: trees which are naturally grown with strong compatibility, stability, and resistance were selected as trees to be preserved. Trees with straight branches, flourishing leaves, no disease, height greater than 0.7 m, and bigger than 0.6 cm in diameter were retained. The trees utilized in the study included sweet gum, red *S. superba*, *A. fortunei*, bitter mesophanerophytes (*Castanopsis sclerophylla*), *Cyclobalanopsis* oak (*Cyclobalanopsis glauca*), white oak (*Quercus fabri*), rock oak (*Lithocarpus glaber*) and Formosan juniper (*Juniperus formosana*). In each hectare, 2220 trees were retained (at the beginning the density was kept at 2850/ha, then 2220/ha were planted 1 year later) as middle layer broad-leaf trees. The retained trees had an average height of 0.75 cm, and an average ground diameter of 1.25 cm. The cutting began during the winter of 1990. After cutting the non-retained trees, a hole of 50 cm × 50 cm × 40 cm was dug for each new tree with row spacing of 3 m × 2 m. In the spring of 1991, high-quality seedlings were selected. Before forestation, the surface soil dug from the hole was returned to the bottom. The roots of seedlings were soaked for about 12 h to moisten the roots with the mud with phosphate fertilizer, in order to keep the soil wet for forestation. The seedlings utilized were straight and firm, with the roots outspread. The retained trees were all shaped, with branches cut off and left on the forest floor to increase soil organic material. Tending was carried out in autumn to create an initial forest with multi-layer trees, shrubs, and grass, which would undergo succession to finally make a forest with the coniferous broadleaf natural community, primarily with conifers.

Reforestation on the barren hills by interplanting trees with secondary broad-leaf brush, step by step, could be done at much greater speed than waiting for natural coverage (Photo 9.2). In that way, it is easier to reach its forest management goal. Compared with the artificial forests, the natural forest has many advantages, such as

Photo 9.2 Closing hills and interplanting trees with secondary broad-leaf brushes step by step for vegetation recovery

time and work saving, low cost, quick effect, and easy forestation. It was also more easily acceptable for people and thus is a good method that is widely applied.

9.4.2 Comprehensive Governing Theory of Small Watershed

The gully region refers to one kind of special region in the barren hills, including the area of slopes and the gully bottom. For a small watershed, the soil and water conditions vary gradually from the top to the bottom. Vegetation is usually scarce on the top of the slope and the uphill, where the soil is thin with low fertility due to soil erosion. But at the bottom the soil is usually thick and has high fertility. Huangbi village of Jingjiang Town, Yujiang County, was selected as an example to introduce the comprehensive governing theory of a small watershed.

1. Designing principle and ideas

   The first principle is to conserve water and soil by planting suitable trees according to local conditions and reasonably selecting the kinds of trees and forests. On the upper parts of the mountain, masson pine, slash pine, red *S. superba*, or other trees with strong resistance to drought and poor fertility were selected. On the middle and downhill parts, loblolly pines and eucalyptus (*Eucalyptus robusta*) were used for afforestation, in order to maintain the water and soil, improve soil fertility, and provide fuelwood. For the stream zone, sweet gum (450/ha) trees were interplanted with *Lespedeza*, white oak, and other trees or shrubs in the coniferous forest. In the downhill parts, Bahia grass and yellow gardenia were planted as grass vegetation to form the multi-layer mixed forests. Terraces at the foot of the slope and the gully bottom were cultivated as orchards primarily and intercropped with peanuts (*Arachis hypogaea*), beans, vegetable, watermelons (*Citrullus lanatus*), or other economic crops. Moreover, in order to do research on the technologies of forestation under this kind of adverse circumstance, different types of forestation techniques, methods, and planting methods were used.
2. Biological measures and the layout

   Biologically comprehensive control measures included forestry, grass, cultivation, and husbandry:

   Forestry: Forestry for small watersheds included planting timber forest, water and soil conservation forest, and fuelwood forest.

   Grass: To restore the vegetation as soon as possible, the hillsides were closed to cultivate grass and integrated with planting Bahia grass among trees, in order to help to conserve the soil and to develop the livestock breeding industry.

   Cultivation: The land in the bottom of stream was cultivated into paddy field, dryland, or small nursery to plant rice and to cultivate the nursery stock, where grain crops were intercropped in the fruit forest and young forest to obtain more food and great economic efficiency.

Husbandry: The forest wild resources and forest by-products were gathered and processed to raise the income of farmers.

Fishing: Some small ponds and reservoirs were excavated in the gully bottoms to develop the fishing industry as well as irrigation. In turn, the pond soil was also used to raise the fertility of the forest.

Chapter 10
Models of Reforestation for Soil Erosion Control in the Hilly Region of the Middle and Lower Reaches of the Yangtze River

Abstract In order to reforest on extremely eroded and degraded red soil sites under harsh conditions, many measures were carried out, such as choosing the appropriate species with strong stress resistance, combining biological and engineering measures, digging 750–1,500 fish scale-shaped pits per hectare and bamboo stem-shaped trenches, and preparing level terrace sites as well as establishing a rational structure and adopting advanced afforestation techniques (ABT rooting powders, using seedling containers, good provenance, high-density and multi-species reforestation, and digging holes for fertilization). One example of the three-dimensional configuration of trees, shrubs, and grasses of the reservoir area in the hilly red soil region showed that in sites of *bicolor Lespedeza* (*Lespedeza bicolor*) intercropped with slash pines (*Pinus elliottii*) after 10 years, the content of organic matter in the soil, total N, and P increased by 87.95, 36, and 40.91%, respectively, compared with the control area. On the hilltops and the upper parts of steep slopes in the reservoir area, forests were planted to conserve soil and water, with slash pine and red pine as main tree species. Meanwhile, measures were adopted by closing the hillsides to use in order to cultivate forests. The "stereoscopic" management model includes the fruit–grass (farming)–animal–biogas–fish model and the fruit–feeds–animal–biogas model. In the low hilly areas, the agroforestry management models were used, such as the so-called fruit–green manure–economic crop model, fruit–grass–economic crop model, the virtuous fruit–*Amaranthus*–pig–biogas–fruit cycle model, tree–tea compound model, and forest–herb management model.

10.1 Introduction

There are some unique advantages of vegetation restoration and reconstruction for the hilly regions in the middle and lower reaches of the Yangtze River. For example, the natural conditions and rich thermal and water resources not only are conducive to the cultivation and development of forest resources in a three-dimensional operating model but also provide a good resource base and options for agroforestry. For barren hills in the middle and lower reaches of the Yangtze River, where secondary broadleaf shrubs are distributed, the forests are planted for wood of high quality by preparing the hills for natural forestation, then replanting the original trees in order to develop community types that not only follow the pattern of natural succession but also meet the objectives of the operation and, in particular, the

J.C. Zhang et al., *Theory and Practice of Soil Loss Control in Eastern China*,
DOI 10.1007/978-1-4419-9679-4_10,

ecological and economic requirements for effective and sustainable forestry. This measure conforms to the requirements of low cost, rapid returns, and high effectiveness in the vegetation restoration and reconstruction of hilly regions. It also lays the material basis for the appropriate handling of the trade-offs between ecological and economic benefits and between short-term and long-term profits.

10.2 The Vegetation Reconstruction Model for Extremely Eroded and Degraded Red Soil Sites Under Harsh Conditions

10.2.1 Site Features

The hilly and mountainous district formed by granite, gneiss, purple shale, and sandstone makes up the largest area of the red soil zone in the south. Subjected to a long period of weathering, this district has a weathering crust with a thickness varying between 10 and 80 m. The surface soil is cemented tightly with high clay content and has poor water permeability. Runoff is formed immediately after rain starts. The red soil, netted with roots, and gravel layers below have poor consolidation, and the soil body breaks as soon as water inside drains away. In addition, the combination of intense precipitation events, soil with great erodibility, and unique undulating topography may lead to water loss and soil erosion. As long as the vegetation is in a damaged state, the surface soil will be washed away by rainstorms and runoff, leading to the deterioration of the nutrient status, the soil structure, and the water content of the soil.

The subtropical red soil region has become one of the most seriously eroded regions of China, owing to its ecological fragility (determined by the characteristics of the soil and climate conditions) and strong human interference. The serious problem of the degradation of the red soil ecosystem, characterized by soil erosion, has caused the deterioration of the ecological environment, constraining the sustainable development of the regional economy.

10.2.2 Guiding Ideologies for Management

Taking vegetation restoration and erosion control as the main objectives, the following work should be carried out to guarantee high survival rates, thus achieving better economic efficiency on the basis of sound ecological practices: constructing a compound forest system with strong ecological functions by rational allocation species of trees, shrubs, and grasses; combining biological measures and measures of soil and water conservation projects; adopting ABT rooting powder, seedling containers, good provenance, and other advanced technologies, as well as forest fertilization, high-density reforestation, etc.

10.2.3 Key Techniques

(1) Choosing the appropriate species with strong stress resistance

The selection of species is not only the basis for vegetation restoration and reconstruction but also a way to control the structure of artificial plant communities. Biological and ecological characteristics of species determine the ecological conditions needed for their normal growth and development. Successful species must be able to survive and propagate themselves across a particular region; thus they must have certain ecological adaptability. Therefore, the selection of species for specific conditions should follow the principle of ecological adaptability.

Due to water loss soil erosion, the soil properties of the red soil zone tend to deteriorate – the soil texture changes suddenly, soil nutrient and water are lost, soil structure degrades, soil acidifies, water storage capacity decreases, and soil development regresses. The site conditions are harsh for afforestation and the growth of trees, and thus appropriate species selection is the key to succeeding in vegetation restoration and reconstruction. Choosing appropriate vegetation with strong stress resistance can guarantee the success in vegetation restoration and reconstruction and secure the stability and safety of the ecosystem. According to a test of vegetation restoration in degraded red soil area of Jiangxi in 1990, slash pine, pond pine *(Pinus serotina)*, red pine (*Pinus massoniana*), loblolly pine (*Pinus taeda*), sweet gum (*Liquidambar formosana*), red *Schima superba* (*Schima wallichii*), river red gum (*Eucalyptus rostrata*), black locust (*Robinia pseudoacacia*), *Ailanthus* (*Ailanthus altissima*), (*Betula luminifera*), white oak (*Quercus fabri*), chen oak, *Lespedeza*, and Bermuda grass (*Cynodon dactylon*) are good for vegetation restoration and reconstruction in seriously eroded red soil areas.

(2) Combining biological and engineering measures

It is hard to carry on afforestation in eroded red soil areas. Besides a reasonable choice of species, appropriate engineering measures, such as retaining water and controlling erosion, are necessary to overcome the difficulties. The soil in the test area in Gan County of Jiangxi is hilly red soil developed from granite with a slope gradient between 10° and 30° and a coverage rate of vegetation between 5 and 10%. There are typical groups of mounds that are easy to level off. During the experiment, we dug 750–1,500 fish scale-shaped pits (Photo 10.1 and Fig. 10.1) per hectare and bamboo stem-shaped trenches, and prepared level terrace sites. In addition, we chose red pine, Tongmian pine, slash pine, sweet gum, red *S. superba*, *Acacia* (*Acacia confuse*) and river red gum as our arboreal species, Lespedeza as shrub species, and broadleaf paspalum (*Paspalum orbiculare*) and crab grass (*Digitaria sanguinalis*) as herbs for the three-dimensional configuration. Because these measures maintained moisture in the soil, the experiment achieved a relatively high survival rate of trees and good ecological benefits.

(3) Establishing a rational structure

"Multi-level matching" is matching the structural features of natural communities, manifested as the diversification of physical and biological structures in time and space, to complement the ecological niche diversity among species, leading to

Photo 10.1 Fish scale-shaped pits

Fig. 10.1 Cross-section of fish scale-shaped pits

stability of communities. Therefore, during vegetation restoration and reconstruction, based on the fact that different species use resources in different places and different times, it is best to maintain the multi-layer feature of soil and water, and match deep-rooted plants with shallow-rooted plants, broadleaf plants with conifers, sciophilous plants with heliophilous plants, deciduous plants with evergreen plants, and trees with shrubs with grasses. A plant community with a rational structure should be built by distributing and composing species rationally, and coordinating intra- and interspecies relationships. As to the severely eroded red soil region, on the one hand, multi-canopy interception and ground surface changes can reduce erosion and runoff, constraining soil and water loss. On the other hand, the net-like structures formed by roots of trees weaving together in shrub–grass–tree system, or systems with various tree species, increase the forest's ability to conserve soil, creating a good environment for the rapid and steady growth of plants. Meanwhile, a good habitat also provides good conditions for the survival and development of other species, improves the biodiversity of the ecosystem, and thus enhances the ecological sustainability and safeguards the ecosystem of the reconstructed forest. For example, in the seriously degraded (grassland) red soil areas of Taihe County in Jiangxi Province, after 10 years' renewal of the forest, the content of organic matter in the soil layer between 0 and 40 cm deep was measured. The measurements showed that the organic matter content varied with different types of forests: pure sweet gum forest > pure red *S. superba* forest > forest of sweet gum interspersed with pines > pure pond pine forest in medium density > forest of sweet gum and pines interspersed by row > pure eucalyptus forest > pure slash pine forest > pure pond pine forest in high density, where the changes were 129, 118, 115, 89, 63, 29, −13, −43%, respectively, compared with the values (6.5 $g \cdot kg^{-1}$) detected before the experiment started in 1990. The total N and the effective N in the soil also follow similar trends.

(4) Adopting advanced afforestation techniques

The combining of advanced afforestation techniques and measures for soil and water conservation is indispensible for restoring vegetative cover in degraded ecosystems. This includes adopting such advanced afforestation techniques as ABT rooting powders, using seedling containers, good provenance, high-density and multi-species reforestation, and digging holes for fertilization. According to the test conducted at the Guifeng experimental site of Yiyang, in Jiangxi, the group treated by ABT rooting powder grew better than did the control group, and the group that underwent container seedling planting performed better than did the bare root seedling group in terms of ground diameter growth and height growth. Compared with the control group, the treatment groups were 16.2 and 11.3% higher, respectively. Since vegetation restoration is related to forest composition, density, and growth, the high-density multi-species composition and fertilization in holes can be said to be good for short-term vegetation restoration and the growth of broadleaf trees. These advanced techniques and engineering measures, together with the multi-layer structure, played an important role in consolidating soil, conserving water, and promoting the survival and growth of vegetation, and thus guaranteed the successful reconstruction of the ecosystem.

10.2.4 Application of Models

(1) Three-dimensional configuration of trees, shrubs, and grasses

Hardly any natural communities are composed of a single species, and most communities are multi-species ones in which a variety of ecological niches exist among species and the different species form a stable community. The understory vegetation plays an essential role in water conservation, nutrient recycling, improving soil structure, and enhancing biodiversity in the forest ecosystem. During vegetation restoration and reconstruction, it is necessary to simulate these characteristics of communities and to impose artificial measures to properly distribute trees, shrubs, and grasses into a three-dimensional structure.

In the extremely degraded red soil areas of Yiyang in Jiangxi, we intercropped *Lespedeza* with slash pines. Ten years later, the content of organic matter in the soil, total N, and P increased by 87.95, 36, and 40.91%, respectively, compared with the control area. The pH value increased by 4.91%. The soil bulk density of the soil layer from the ground surface to 40 cm deep decreased by 13.2% on average. The capillary water capacity, the saturated water content, and the total porosity also significantly increased by 37.5, 54.86, and 42.31%, respectively, and the permeability coefficient increased by 145.4%. All these figures showed that compaction, ventilation, water retention, and saturated infiltration of the degraded red soil have been improved effectively. In addition, the activity of soil protease, catalase, soil invertase, and acid phosphatase in the 0–40 cm layer was generally higher than that of the control area, which further showed that the 10 years' three-dimensional distribution model of slash pine × *Lespedeza* effectively improved the decomposition of N from protein in the soil, increased the maturity of the soil, effectively raised the available P in soil, and increased the ability of the rate of decomposition of toxic substances in the soil, all conducive to the maintenance of soil fertility.

(2) Multi-species mixed model

The multi-species mixed model has proven to be a successful model for vegetation restoration in degraded red soil areas. According to this model, drought and infertility-resistant species are selected and the principle of making full use of the resources and maximizing ecological and economic benefits is adopted.

In 2002, we adopted the mixed-plant models in the degraded red soil areas of Zhukeng of Yiyang in Jiangxi (within which our experimental zones were scattered). These four mixed models included slash pine × sweet gum × red *S. superba*, *Elaeocarpus decipiens* (*Elaeocarpus sylvestris*) × Chinese tulip tree (*Liriodendron chinense*) × red *S. superba*, red *S. superba* × Chinaberry (*Melia azedarach*) × Betula luminifera, *Ailanthus* × black locust × river red gum. *Lespedeza* was interplanted in all of these models. Three-year test results show that soil bulk densities of the four models were lower than those of the control group by 5.57, 3.96, 11.32, and 8.41%. The conditions of all the nutrients of the soil have been improved, except for effective K.

10.3 The Stereoscopic Management Model for the Reservoir Area in the Hilly Red Soil Region

The major soil type of the large upland areas in the middle and lower reaches of the Yangtze River is red soil; it is dry, infertile, acid, sticky, compact, corroded, and weak in its ability to hold fertilizer and water. Meanwhile, because of the commonly uneven distribution of rainfall, droughts occur frequently in this hilly red soil region. An effective way to regulate water in this region is to build reservoirs. In fact a lot of reservoirs have been developed here. For example, Nancheng, a county in Jiangxi Province, has more than 60 small reservoirs. Yet, most of the reservoir areas are in poor condition: soil and water are being lost due to the practice of emphasizing agriculture over forestry, caused by the lack of sustainable development of modern agriculture. The planting structure is not appropriate and the emphasis is on rice production, to the neglect of the development of other economic products; land capacity is declining due to the overuse and poor care of the soil, causing a vicious cycle in the small watershed ecosystems. Therefore, we should find an effective means for the development of reservoir areas in hilly red soil region by treating the ecological system of the reservoir and surrounding area as a unit and properly designing and allocating the elements, functions, and structures according to the stereoscopic management model.

10.3.1 Elements of the Design

The model includes the elements of forest, orchard, grass (*Amaranthus*), animal husbandry, agriculture, biogas, and fishery:

Forest: On the top and the upper parts of steep slopes in the reservoir area, we can build forests to conserve soil and water, with slash pine and red pine as main tree species. Meanwhile, measures should be adopted by closing the hillsides (Photos 10.2 and 10.3) in order to cultivate forests.

Orchard: In the gently sloping-terrain fields, it is best to plant fruits like Nanfeng oranges, Wenzhou oranges, chestnuts, grapes, peaches, and southern apples, making them part of the major contents for the three-dimensional development.

Grain: In the original farmland and the gently sloping areas of the reservoir shore zone, rice should be planted. In unoccupied places of orchards and

(a) (b)

Photo 10.2 Closing the hillsides in order to cultivate forests in China. (**a**) Local residents enclosing hillsides with iron wire to protect it from pasture and gathering of firewood and wood for 10 years; (**b**) a warning sign board with Chinese characters stipulating that no fires should be made within the enclosing hillsides

Photo 10.3 Enclosing the rocky mountain after tree plantation in Xuzhou, Jiangsu province, China (2010.9)

ramp drylands, upland rice, corn, etc. should be planted, which can provide not only rice for human consumption but also livestock feeds.

Economic crops: In the unoccupied places within orchards and beside reservoirs, peanuts (*Arachis hypogaea*), soybeans (*Glycine max*), cotton (*Gossypium herbaceum*), beans (*Vicia faba*), and other vegetables should be planted, which can bring short-term economic benefits. Meanwhile, soil can be improved, land fertility maintained, and sustainable development promoted because of large number of residual leaves and rhizobium.

Grass: When a new orchard is cultivated, full or strip tillage is often adopted, which completely removes the covering materials, and thus damages the original soil structure, resulting in more vacant areas in the orchard and aggravating soil erosion. Planting grasses in unoccupied places is an effective way to improve ground coverage and control the erosion. Bahia grass (*Paspalum notatum*) could be interplanted in the orchard or on slopes to protect water and soil. In addition, as we know, grain amaranth (*Amaranthus hypochondriacus*) has a strong ability to absorb potassium. It could be planted in unoccupied places of the orchard to obtain low-cost and high-quality feeds, absorb and activate the non-exchangeable potassium in the soil, and also to effectively promote the circulation of the soil in the orchard and maintain the balance of soil nutrient by making green manure or food for poultry from its stems.

Animal husbandry: Animal husbandry should be developed with pigs as the main livestock and poultry as an auxiliary livestock in order to guarantee the short-term benefits of the orchards and provide orchards with adequate sources of organic fertilizer, and thus achieve sustainable development of orchards.

Biogas: In order to make full use of animal manure and garbage to achieve the multi-level circulation of materials and enhance comprehensive benefits, it is best to establish a livestock farm and biogas digesters, through which animal manure can be used to produce methane to provide energy in winter and biogas fluid, which can be used as fish food.

Fish: Reservoirs are the basis for the economic development in reservoir areas. This fact argues that fish culture should be the main subject of development. Full use should be made of the water resource and to increase the profits of fish culture by choosing proper fish species and adopting mixed and intensive culture techniques. Bighead carp, silver carp, grass carp, and carp fingerlings should be chosen as major species, together with California perch, bream, freshwater white butterfish, and other valuable fish.

10.3.2 Stereoscopic Management Model

We chose two orchards in Tianjingyuan, a town in Nancheng County of Jiangxi Province, as the experimental subjects for the construction of the main management model for the reservoir area. Covering an area of 20 hm^2, the reservoir is located in a typical low hilly region at an elevation of 148 m. The soil is red sandy soil developed from red sandstone and purple shale, and its pH value is between 5 and 6. It has good permeability and poor water- and fertilizer-holding capacity.

(1) Fruit–grass (farming)–animal–biogas–fish model
The orchards on the banks of the reservoir were taken as the management unit and the water in the reservoir as the management base (covering an area of 20 ha). The gentle slopes beside the reservoir were developed to plant Nanfeng orange trees (2 ha) and rice (0.27 ha) and unoccupied places of the orchard were used to grow grain amaranth (0.4 ha), together with Bahia grass (0.13 ha). Six hundred pigs and 600 chickens can be raised by feeding them grain amaranth. Biogas could be made from animal manure and biogas fluid, which is used in fish culture, can be used to result in harvests of 1,500 kg fish per year, as well as the raising of 3000 ducks and 60 geese on the water surface. The mud in the bottom of the reservoir could be used to fertilize fruit trees and crops. In this way, a positive cycle has been formed, in which different species grow together. Long-, medium-, and short-term benefits are combined, and ecological, economic, and social benefits are increasing together. A schematic of the material circulation of the model system is shown in Fig. 10.2.

(2) Fruit–feeds–animal–biogas model
An orchard in the hillocks of the reservoir area was taken as an operating unit. Based on the 13.3 ha of orchards, full use was made of the unoccupied spaces to introduce feed crops such as grain amaranth, corn, upland rice of Brazil, and economic crops such as peanuts, watermelons (*Citrullus lanatus*), beans, pea (*Pisum sativum*), and cotton. Pigs were fed with low-cost feeds made of grain amaranth (green fodder) (Photo 10.4), dry leaves and stems, and corn and cereal powder. Five hundred and seventy-two pigs could be raised per year, and pig manure could be used to produce biogas and biogas fluid. Methane could provide energy in winter and biogas fluid could be used as organic fertilizer. In this model, the planting structure has been rearranged by growing good feed crops, which then guarantees the possibility of low-cost animal culture, which is a breakthrough in this system, promoting the stable functioning of the pig–biogas–fruit project. Material circulations of this model are shown in Fig. 10.3.

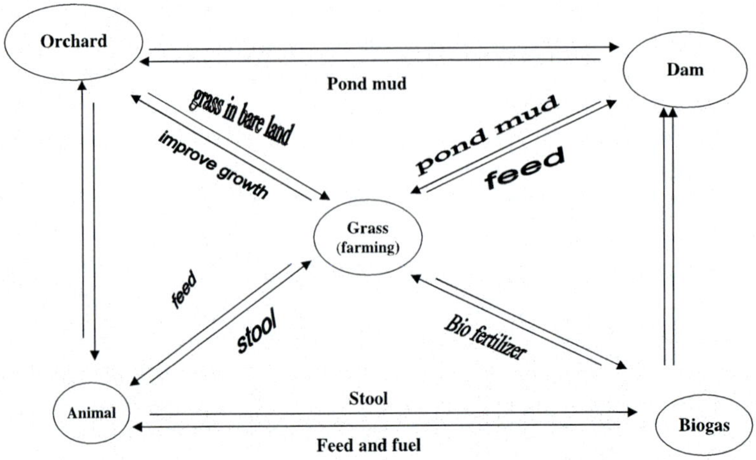

Fig. 10.2 Diagram of material circulation

Photo 10.4 Feeding pigs with low-cost feeds made of grain amaranth

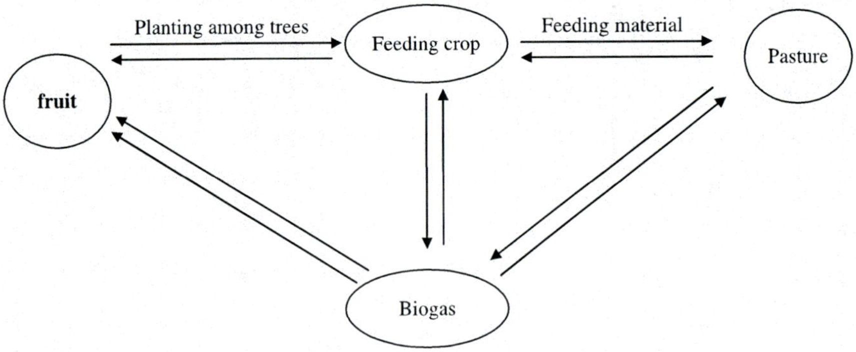

Fig. 10.3 Material circulations of fruit-feeds-animal-biogas model

10.4 The Vegetation Restoration Model for Harsh Limestone Areas

10.4.1 The Features of the Harsh Limestone Areas

The climate of the middle and lower reaches of the Yangtze River is characterized by abundant heat and rainfall and a synchronization of rainfall and temperature. These characteristics of climate, together with thin soil layer on slopes and the uneven distribution of soil in cracks and on rock surface, result in the uneven distribution and the overall low coverage of vegetation. Moreover, arable lands in this limestone mountainous area are rare. Intensive development of slope lands has accelerated the soil erosion. Therefore, these ecologically fragile areas are difficult for vegetation restoration.

Jiujiang is a city in the middle and lower reaches of the Yangtze River, which has the most extensive area of limestone. Its limestone mountainous area accounts for 45.9% of the total limestone mountainous area in Jiangxi Province. The parent materials of the soil are mainly composed of calcium carbonate limestone, argillaceous limestone, and quartz-based siliceous limestone. There are many common features between them. First, they are soluble and developed from fissure caves. Second, the soil layer is thin but rich in calcium and organic matter. Third, the soil is either neutral or slightly acidic or slightly alkaline. Fourth, it is hard for these soils to keep water and fertilizer and the ground surface is apt to be dry.

10.4.2 The Guiding Ideology for Management

Although there are many difficulties for vegetation restoration in limestone areas, in certain layers, certain parts of the soil layers are fertile, nearly neutral in pH, and good in texture. Therefore, the main management objective for these areas is to combine management with development and utilization to establish a management model emphasizing both economic and ecological benefits. One should select and stereoscopically combine the right tree and shrub species which not only are adapted to limestone areas but also can have definite economic value. The idea is to improve the survival and preservation rate of the plants and, at the same time, organically combine ecological and economic benefits and improve the local economy.

10.4.3 Key Techniques

(1) Restoration according to different site conditions

There are many limiting factors for vegetation restoration in harsh limestone areas. These factors are different in their influence at different spatial levels and scales.

At the large scale, there are rainfall, temperature, and other climatic factors, as well as the nature of the rock. These factors, in the macro sense, influence the type of the site and quality of the soil and the degree of difficulty of reforestation. At the small scale, the factors include elevation, slope, aspect, position of the slope, and the exposed fraction of rocks. These factors affect the thickness, texture, quality, moisture, and other factors of soil and vegetation conditions, and thus determine the degree of difficulty of vegetation restoration to certain extent. Therefore, to succeed in afforestation, it is essential to evaluate and classify the conditions of harsh sites, analyze the features of communities, and select the right species.

Based on the degree of difficulty of the site and factors affecting restoration (the bare rock fraction, slope, and coverage degree of vegetation), we divided Ruichang, the major limestone area of Jiangxi, and the Pingxiang limestone areas into four groups by using GIS graphics processing software. Site types and features of the communities of each site group are as follows:

Class I. It is the most difficult site type. This type is usually distributed on slopes with a gradient of over 25° and a bare rock fraction of more than 70%. Its soil has an average thickness of 10 cm and is very low in nutrients. The community physiognomy consists of sparse rattan shrubs, of which Chinese raspberry (*Rubus parvifolius*, Photo 10.5) is absolutely the dominant species with an importance value of 52.9%. Following it are bitter bamboo, sweetbrier, and Chinese *Rubus* (*Rubus saxatilis*, Photo 10.6). There are only 13 species of shrubs.

Class II. It is the second most difficult site type. This type of forest is usually scattered on various slopes that have a bare rock fraction of more than 50%, and the thickness of the soil layer is between 20 and 40 cm. It has a little more nutrient compared to Class I. The community physiognomy is sparse rattan shrubs or shrub meadow, together with scattered trees. The dominant species are sweetbrier, small fruit rose (*Rosa cymosa*), smooth-leaved *Smilax*, and some other echinate species, of which the importance value of species

Photo 10.5 Chinese raspberry with fruits

Photo 10.6 Chinese *Rubus* with fruits

decreases visibly and sweetbrier accounts for 28.3%, the highest importance value. The total number of shrub species reaches 34.

Class III. It is the slightly difficult site type. This type is usually located in places with a gradient of over 25°, a bare rock fraction of more than 30%, and relatively high nutrient level. The average thickness of its soil layer is 40 cm. The community physiognomy is shrub meadow, together with scattered trees, sweetbrier, small fruit rose, smooth-leaved *Smilax*, and some other echinate species, of which the importance value of further species decreases and small fruit rose has the highest importance value, which being only 23.5%. Meanwhile, Chinese *Loropetalum* (*Loropetalum chinensis*), white oak, and other woody shrubs have increased and there are a total of 48 kinds of shrub species.

Class IV. This site type is relatively easy to restore. This type is usually distributed in regions with a gradient of less than 25°, a bare rock fraction of less than 30%, and a relatively high nutrient level. The average thickness of its soil layer is 60 cm. The community physiognomy is grass slopes covered with scattered trees. The importance value of June snow (*Serissa japonica*), glutinous rice (*Oryza sativa*), white oak, Chinese Loropetalum (*L. chinensis*), mountain pepper (*Lindera glauca*), and other woody plants is higher in this type.

(2) Using phosphate fertilizer properly

Because phosphorus often exists in the form of insoluble phosphorite in the soil of the limestone areas, causing a lack of available phosphorus, a proper amount of phosphate fertilizer should be used during afforestation.

(3) Selecting appropriate species

Soil developed from limestone is rich in Ca and Mg. The exposed area of rocks is large and it is easy for drought conditions to occur. Therefore, many tree species are in poor growing conditions, leading to low survival rate. Yet, compared with red soil in other places, it has its advantages – it has low acidity, is almost neutral in pH, and has relatively high soil fertility. During the choosing of tree species, many factors should be considered, including rapid growth, benefits to the maintaining of soil fertility, and economic values. In addition, special consideration should be paid to the unique features of soil developed from limestone. According to the experiments in the limestone mountainous areas of Xiushui and Ruichang in Jiangxi Province, some arboreal species (Chinese arborvitae (*Platycladus orientalis*), Sichuan alders (*Alnus formosana*), *Cedrela sinensis* (*Toona sinensis*), *Eucommia* bark (*Eucommia ulmoides*), phoenix tree (*Firmiana simplex*) (Photo 10.7), dogwood (Co*rnus officinalis*), Wheel wingnut (*Cyclocarya paliurus*), black locust, Sapindaceae, Chinese tallow tree (*Sapium sebiferum*), goldenrain tree (*Koelreuteria paniculata*), *Ailanthus*, palm trees (*Trachycarpus fortunei*), etc.), shrub species *Amorpha* (*Amorpha fruticosa*), *Lespedeza*, rutaecarpine (*Tetradium ruticarpum*), lacquer tree (*Toxicodendron vernicifluum*) (Photo 10.8) etc.) and pale bamboo (*Phyllostachys glauca*) have good adaptability. These species can bring in good ecological and economic benefits and thus are suitable species for vegetation restoration in limestone areas.

(4) Selecting appropriate places and times for afforestation

Difficult limestone areas have large areas of exposed rocks, which absorb heat rapidly, causing quick evaporation of soil moisture. Besides, there are many rock

Photo 10.7 Chinese phoenix tree

Photo 10.8 Chinese lacquer tree

caves and underground holes, leading to water leakage. Therefore, there is a shortage of moisture on the surface of and even inside the soil, and consequently there is weak drought resistance capability. During afforestation, special attention should be paid to choosing the appropriate places, season, and time, and to making full use of microtopography with appropriate thickness of the soil layer in making the tree-planting pits. Rainy spring days are good for afforestation. Cover grass around young trees during cultivation keeps soil moisture and increases the survival and the preservation rates.

10.4.4 Application of Models

(1) Coniferous–broadleaf mixed model

In 1991, we mixed Chinese arborvitae with phoenix trees (Photo 10.7) (with a mixture ratio of 6:4) and Chinese arborvitae with *C. sinensis* (with a mixture ratio of 9:1) for the restoration in Xiushui County of Jiangxi Province. According to investigations in 1994 and 2003, soil bulk density decreased visibly. Moreover, soils carrying 12-year-old trees were greatly improved in terms of water-holding capacity and capillary interstices relative to soils carrying 4-year-old trees. Compared with pure *C. sinensis* and pure sweet gum forests of the same age, Chinese arborvitae–phoenix tree and Chinese arborvitae–*C. sinensis* mixed forests had better soil bulk density, capillary water-holding capability and capillary interstices, and thus were better for the improvement of water-holding capability and the capillary interstices of the soil. In addition, the nutrient conditions had been improved to some extent, especially the content of available nutrients and the base cation exchange capacity. These facts show that these models play an important role in maintaining and improving soil fertility.

(2) Mixed broadleaf model

Based on the biological and ecological characteristics of trees, various broadleaf trees could be mixed in order to fully utilize resources and achieve the largest ecological and economic benefits. Experiments in Xiushui and Ruichang of Jiangxi Province showed that many mixed models had excellent results. They were Chinese aralis (*Cornus macrophylla*) plus *C. sinensis* (Photo 10.9), Chinese aralis plus *Amorpha* (*A. fruticosa*), Sichuan alders plus sweet gum plus phoenix tree, *Eucommia* bark plus *C. sinensis* plus soapnut tree (*Sapindus mukorossi*), Azadirachta indica plus black locust plus *Ailanthus*, *Eucommia* bark plus *C. sinensis* plus goldenrain tree (*K. paniculata*), *C. sinensis* plus Sichuan alders plus Chinese scholar tree, *Ailanthus* plus black locust and soapnut tree, *C. sinensis* plus *Ailanthus* and Chinese tallow tree, and Sichuan alders plus *Eucommia* bark plus *Ailanthus*.

The model of Chinese aralis plus *C. sinensis* in degraded barren hills of Xiushui in Jiangxi Province is presented as an example.

Four years after reforestation (Photo 10.9b), the organic matter, total N and P, available N, P, and K, and base exchange capacity in the soil layer between ground surface and 40 cm deep inside increased to various degrees. The base exchange capacity increased to 125.9 mmol kg^{-1}, compared with 36.09 mmol kg^{-1} before afforestation. The content of organic matter in soil increased from 2.05 to 3.15% and the available N increased from 84.34 to 109.81 g kg^{-1}. In the mixed model of Sichuan alders, sweet gum, and phoenix tree, the average diameter at breast height of the 12-year-old Sichuan alders, sweet gums, and phoenix trees reached 12.0, 58.60, and 10.07 cm, respectively, and the average height reached 12.34, 7.90, and 10.50 m, respectively. The total N, P, K, available N, available P and K, base exchange capacity reached 0.165%, 0.0412%, 0.98%, 110.06 g kg^{-1}, 3.94 g kg^{-1}, 42.95 g kg^{-1}, and 57.65 mmol kg^{-1}, respectively. This model resulted in the highest level of total N and available N among all test models, which indicates that it functions better than did the others in maintaining soil fertility. In the mixed configuration of 12-year-old alders, sweet gums, and phoenix trees, the bulk density, capillary water-holding capacity, saturated water content, capillary porosity, non-capillary porosity, and the total porosity of the soil between ground surface and 40 cm deep are 19 g cm^{-3}, 45.3, 47.6, 52.3, 2.8, and 55.1%, respectively, which indicate that this model functions well in improving the physicochemical properties of soil and maintaining water and fertilizer.

(3) The forest medicine management model

Two species fit in this model – Fructus Evodiae (*Evodia rutaecarpa*) and Yellow gardenia (*Gardenia stenophylla*). Fructus Evodiae is a kind of small tree belonging to Rutaceae family. Its fruits, leaves, roots, stems, and skin can be used as medicine. Its young fruits can be used effectively to treat heartache, abdominal pains, and stomach pains. Distributed in provinces to the south of the Yangtze River, it can adapt to various soil types, except the overly sticky and heavy types. It can grow in neutral, slightly acidic, and slightly alkaline soil and prefers sandy, oil sandy soil, and other soil types with adequate fertilization, moisture, and good drainage. Yellow gardenia is a kind of shrub belonging to the madder family. Its fruits can be used as

Photo 10.9 Mixed model of Chinese aralis plus *C. sinensis* in Xiushui prefecture of Jiangxi Province, China. Bare hill (**a**) before reforestation in 1991; (**b**) 4 years after reforestation in 1995; (**c**) 13 years after reforestation in 2003

(a)

(b)

(c)

natural yellow dye. It usually grows in low-elevation forests in the region south of the Yangtze River or in southwest China, prefers warm and humid climates, and has such good features as cold, drought, and pruning resistance. It grows well in fertile, moist, well-drained acidic soil or neutral and slightly acidic red soil.

Fructus Evodiae–Yellow gardenia or Fructus Evodiae–*Eucommia* bark model (mixed row by row) can be adopted in those areas with relatively good site conditions. These models were applied in Ruichang City and Pingxiang of Jiangxi Province and resulted in a 94% survival rate of Fructus Evodiae and 95% survival rate of Yellow gardenia. The average height, ground diameter, and crown breadth of 2-year-old Fructus Evodiae were 181, 5.3, and 176 cm, respectively. The average diameter, height, and crown breadth of 2-year-old Yellow gardenia were 1.27, 71, and 60 cm, respectively. The woodland was basically at the closed canopy stage and the soil erosion condition was under control.

(4) The mixed model of pale bamboo and broad-leaf trees

Pale bamboo McClure, a kind of calcium-tolerant plant adapted to limestone soil, is an important economic bamboo species in limestone mountainous areas. The planting of pale bamboo not only can promote the development of the bamboo industry and bring in more profits for the local residents but also plays an important role in soil and water conservation, improving the local ecological environment. No other species can surpass it in these respects. In the natural conditions of this region, this model produces a relatively stable ecosystem and strong capability in maintaining soil, fertility, and water storage. For example, in the limestone areas of Ruichang City in Jiangxi, pale bamboo–camphor tree and pale bamboo–sweet gum models were applied with camphor trees and sweet gum trees irregularly interplanted in pale bamboo. In these models, pale bamboo and broadleaf trees were in proportion of 8:2. The satisfactory ecological and economic benefits achieved indicate that the mixed model of pale bamboo and broad-leaf trees is good for vegetation restoration in limestone areas.

10.5 Vegetation Restoration Models in the Abandoned Mining Areas

10.5.1 Site Features

The middle and lower reaches of the Yangtze River are rich in mineral resources. As we know, during mining exploitation processes, no matter whether it is on the surface or in deep layers, a large quantity of soil is moved away or buried by mineral wastes. As a result, the ecosystem has been damaged, or even severely ruined. The extreme soil conditions (poor physical conditions, the lack of nutrients, and toxicity owing to abnormal pH value and heavy metals) would then hinder the growth of plants. Therefore, it is difficult to carry out vegetation restoration in the abandoned mining areas.

10.5.2 Guiding Ideology for Management

The major objective of vegetation restoration and reconstruction in these sites is to restore and reconstruct the forest and vegetation system on the abandoned mining areas and improve the ecological environment. Based on the analysis of the obstacles for the growth of plants in the mining areas, appropriate plant species (particularly those that have ameliorating effects in the polluted areas) can be selected and combined with biological and engineering measures to rebuild a rational plant community.

10.5.3 Key Techniques

(1) Selecting suitable plant species

Suitable plant species can be selected and combined on the basis of studies of community composition, quantitative properties, the interspecific relationships, and floristic characteristics. The rules of regional distribution, dominant species, and composition of natural communities can be applied. For example, from the dynamic vegetation and community investigation on Mufu Hill in Nanjing, it was learned that, due to human interference, there is no complete natural zonal vegetation system left in Mufu Hill, which is a part of the north subtropical, evergreen, broadleaf deciduous forest. The existing vegetation communities are basically secondary, mainly composed of common paper mulberry (*Broussonetia papyrifera*), chikrasi (*Chukrasia tabularis*), black locust, hackberry (*Celtis sinensis*), etc. The tree layer consists of such arboreal species as common paper mulberry, hackberry, Amur maple (*Acer ginnala*), live oak (*Quercus dentata*), Chinese mulberry (*Cudrania tricuspidata*), dye tree (*Platycarya strobilacea*), and white elm (*Ulmus pumila*), and such understory species as *Alangium* (*Alangium chinense*), Chinese wild pepper (*Zanthoxylum planispinum*), Cascara (*Rhamnus utilis*), *Symplocos* (*Symplocos lancifolia*), and mulberry.

The coexisting species in the communities can be separated into two types of dynamic trends in the vertical structure. One type includes the invasive species (paper mulberry and hackberry) and the other includes the degenerate species (black locust). The diameter distribution of individual trees is narrow and more trees are in the lower ranges, which indicates that the community has good capacity for undergoing succession. All these facts have provided important reference points for vegetation restoration for the abandoned mining area in Mufu Hill (Photo 10.10).

Photo 10.10 shows tractors compacting soil, which was carried by trucks from the tunnel excavation work for Nanjing's underground to Mufu Hill, Nanjing, China. For the abandoned mining area in Mufu Hill, utilization of the excavated soil facilitates afforestation.

(2) Vegetation restoration techniques with multiple targets

Deciding on appropriate targets is very important for ecological restoration. There are various kinds of degraded mining areas, such as areas covered by waste stone and mud dumped during exploitation, areas of abandoned slag, and collapsed

(a) (b)

Photo 10.10 Vegetation restoration for the abandoned mining area in Mufu Hill, Nanjing, China: (**a**) before vegetation restoration; (**b**) after vegetation restoration

places caused by exploitation. Therefore, it is important to choose different types of targets for the vegetation restoration, based on the degradation levels of the ecological environment and soil, and for the role this system will play in the future and in relation to surrounding landscapes. Therefore, vegetation restoration with multiple targets for the abandoned mining areas is recommended. In a word, the vegetation restoration and ecological reconstruction of abandoned mining areas may have multiple targets, which could include ecological, scenic, and timber forest goals.

The construction of ecological parks is one method of vegetation restoration and reconstruction of abandoned mining areas. A natural, effective, stable, and economic structure of green land could be built if we design, construct, and manage it based on the theories of ecology and landscape ecology, make use of the structure and succession process of natural vegetation, and build various habitat types with local features based on the differences of natural environments (mainly in water and soil).

(3) Making limiting factors as dominant factors

The minerals in the middle and lower reaches of the Yangtze River differ in types, utilization approaches, and mining methods, resulting in a variety of different site conditions. Some areas are severely damaged on the surface, some are contaminated by heavy metals, some have serious soil and water loss, and some have very harsh site conditions. In other words, the factors influencing vegetation restoration are various. During vegetation restoration, we should make the limiting factors the dominant factors for consideration. For areas with serious soil and water loss, one should combine biological and engineering measures. For areas contaminated by heavy metals, one should choose heavy metal-resistant plant species that can take up and concentrate heavy metals. For areas seriously damaged on the surface, one should choose and combine proper plant species to protect the ground surface.

(4) Adopting the composite restoration model with trees, shrubs, and grass

Research shows that a good method of building a compound structure is to mix woody plants and herbaceous plant that have the ability to absorb heavy metals to purify soil in heavily polluted areas. The effect is obvious, the method easy, and economic benefits high.

Take the study in Dai Village (in the upstream reaches of the Le'an River in Leping City of Jiangxi Province), for example. Four kinds of trees with strong tolerance ability to heavy metals and three kinds of herbaceous plants with strong ability to absorb heavy metals were selected and allocated for planting. Three models were formed: poplar, copper flower, *Rhododendron*; bur beggarticks herb (*Bidens tripartita*) and red maple (*Acer rubrum*); and truncate maple (*Acer truncatum*) and acid grass (*Oxalis corniculata*). The results show that copper flower and acid grass are good at absorbing Cu and Zn, while Bur Beggarticks herb is good at absorbing Cu and Cd. After a period of time, the metal-concentrating plants were removed away and either buried as wastes or the heavy metals were extracted from them. In this way, the polluted soil was also utilized during restoration, creating environmental, economic, and social benefits. It is an absolutely new approach of getting rid of the combined pollution of Cu and Cd.

10.6 The Agroforestry Management Models

10.6.1 The Agroforestry Management Models in Low Hilly Areas

Unique advantages in the natural conditions, climate, and rich water resources make the mountainous region in the middle and lower reaches of the Yangtze River a suitable area for the cultivation and development of forest resources and the construction of the stereoscopic management model. Soil on level and slightly sloping areas is thick and relatively fertile, and thus is reclaimed into horizontal strips around hills for detailed management. However, there are also some disadvantages, such as soft soil on some slopes, low planting density, large spaces between woods, and poor drought resistance. In recent years, because of the large-scale rapid development of the forestry and fruit industries in mountainous regions, many forestlands have been changed into orchards. As a result, many problems arise due to such factors as climate, rainfall, mother rock, and inappropriate use – including the damage of vegetation, low land utilization rate of new orchards, shortage of covering vegetation, and serious soil and water loss. Although the farmers have made active use of stereoscopic management models, comprehensive benefits have not been given full play owing to the lack of guiding techniques.

10.6.1.1 Guiding Ideology for Management

A fruit industry system can be built that combines ecological restoration and economic development. The structure of the ecosystem of orchards can be regulated based on the following two principles: the principle of obeying interspecies complementarity, distinct gradations, and multi-layer utilization, and the principle of combining planting and protecting, long-term and short-term benefits, and sustainable development. Meanwhile, the theory of energy circulation and material transformation can be adopted to guide the practice of the fruit industry and to use suitable stereoscopic management models according to different phases and characteristics of the orchards.

10.6.1.2 Key Techniques

(1) Constructing various models according to different management objectives

One can move away from the traditional monoculture model by interplanting green manure or economic crops in suitable places and by utilizing the temporal and spatial differences in growth among different species. Various stereoscopic management models thus come into being. Habitat diversity was created by making use of the functions of multi-level, multi-pattern, and multi-species plant communities. These were planted to optimize the structure of species groups in the orchards, and consequently to improve the utilization rate of the space and land of the orchards and maximize the use of resources and bring more profits for the orchards.

(2) Designing management models based on the principle of ecological benefits first

The concrete measures were as follows: increase the area covered by vegetation, ensuring a whole-year vegetation cover in the orchards; improve temperature, humidity, and other micrometeorological conditions of the orchards; regulate the soil moisture condition in the drought and flooding seasons, and limit soil and water loss; and improve soil structure, increase the content of organic matter in the soil and water storage capacity, and thus maintain soil fertility and promote the growth of fruits.

(3) Protecting the slopes with grass and building biological fences (such as ditches along the hillside)

By doing this, a water-conserving ecological barrier would be formed, through which water and soil are conserved, promoting the transformation of the environment from harsh to positive.

(4) Attaching importance to the complementary relationships among biological populations in the orchards

The comprehensive treatment of orchards is a complex ecological project. In the choice of types and directions of management, one should adhere to the theories of ecology and economics and build a multi-component structure, with the fruit industry as the main element and other biological populations as auxiliary elements. Attention should be paid not only to the productivity of fruit trees but also to the maintenance of soil fertility, environmental protection, and sustainable production, during choosing the types and managing techniques, to achieve the ecological benefits along with the economic benefits.

(5) Achieving a virtuous cycle of the ecological economy by combining planting and livestock breeding.

Make use of the important roles pigs, cattle, and small livestock play in maintaining the circulation of nutrients and soil fertility in strongly leached soil. Provide the material foundation for the livestock raising industry by planting fodder, green manure crops, and fresh grass. Particularly, use livestock manure to produce methane, which can not only provide energy for domestic use and lighting but also reduce environmental pollution.

Meanwhile, returning biogas fluid and residue to the orchards can effectively prevent soil degradation, improve soil fertility, and reduce the incidence of pests and diseases.

10.6.1.3 Application of Models

(1) Fruit–green manure–economic crop model

This model combines green manure crops and the traditional compound management techniques. Its main objectives are to support long-term benefits with short-term ones, replace cultivation with cropping, and improve soil fertility. In other words, the idea is to arrange green manure and economic crops reasonably. This involves paying attention to the complementary and mutually beneficial relationships that exist among species populations over time and space scales, and their effects on improving soil and conserving water, based on the conditions of soil erosion. It also involves combining the orchards with different species groups according to their stages of development, their characteristics, and the management requirements, achieving an overall vegetation cover in the orchards. This model is suitable for new orchards with low fertility and severe soil erosion.

The specific measures are as follows: for new orchards with low fertility, green manure crops of 1–2 applications per year are indispensable. One can interplant radish (*Raphanus sativus*), milk vetch (*Astragalus sinicus*), pigeon pea (*Cajanus cajan*), *Crotalaria* (*Crotalaria pallida*), common cowpea (*Vigna unguiculata*), and other species with strong tolerance to dry and barren conditions to improve soil and conserve water. For orchards with better site conditions and relatively improved soil, one can grow green manure crops in autumn and winter and economic crops in summer and spring, and in the vacant places between rows of fruit trees, short economic crops such as peanuts, soybeans, watermelons, potatoes (*Solanum tuberosum*), and tobacco (*Nicotiana tabacum*) can be planted. This method not only increases the ground coverage in the orchard, reduces erosion and soil temperature in summer, maintains soil temperature in winter, improves the physical and chemical properties of soil, and improves soil fertility but also brings in an assured level of profit, realizing the goal of nurturing the orchard by making use of its own resources. Consider Lai village, Tiantou, Changsheng, and some other places in Ningdu County of Jiangxi Province, for example. In orchards with fruit trees planted for a relatively long period of time and partly matured soil, we interplanted radish in winter and economic crops (peanuts, watermelons, beans, and vegetables) in spring, summer, and autumn. However, in the barren land planted with 1-year-old fruit trees, we planted pigeon pea, *Crotalaria*, and radish and other green manure crops and began to interplant economic crops in the second year. In this way, we not only improved soil quality, conserved soil and water, and covered the orchard with vegetation during the whole year but also realized a multi-layer utilization of soil fertility in time and space and achieved considerable economic benefits.

(2) Fruit–grass (Bahia grass)–economic crop model

This model is generally utilized in new orchards with relatively large elevation gradients. The specific steps are as follows: Introduce Bahia grass or cultivate the original vegetation to fix soil and prevent water erosion; interplant and intercrop economic crops to form a fruit (navel orange (*Citrus reticulate*))–grass (Bahia grass)–economic crop model; and thus improve the micrometeorological conditions in the orchards and increase economic benefits (Photos 10.11 and 10.12).

Bahia grass (Photo 10.13) is a kind of perennial species belonging to the Poaceae family. It has strong drug potency, a widespread root system, and has drought, cold, barren-land, and trampling-tolerant features, which guarantee its strong ability to retain soil. Besides, it is a good animal feed, as its young grass shoots are tender and full of nutrition. It was initially introduced from Taiwan in 1989 and cultivated in southern Jiangxi Province in two ways (as shown in Fig. 10.4), i.e., it was planted on slope walls between fruit belts with large gradients or along hill ditches to form biological cover for slopes in order to conserve soil and water. Hill ditches were trapezoidally shaped cross-sections designed to retain runoff. Along the contour intervals in the orchard, counter-sloping sections were set up with inner parts lower than outer ones and the plant Bahia grass planted on the surface, forming

Photo 10.11 Naval orange tree orchard on sloping land in southern Jiangxi Province

Photo 10.12 Naval orange tree orchard garden on sloping land in southern Jiangxi Province

Photo 10.13 Bahia grass

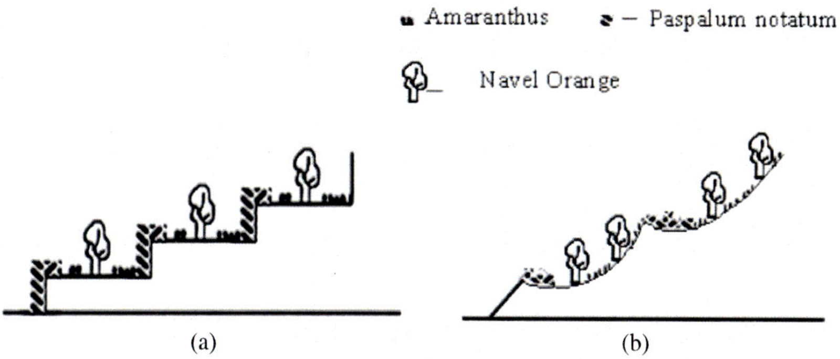

Fig. 10.4 Planting of Bahia grass on slope walls or along hillside ditches. (**a**) Bahia grass–economic and green manure crops on slope walls; (**b**) Bahia grass–economic crops along hillside ditches

a biological fence with fruit trees and grass strips. Meanwhile, economic crops in the vacant space between fruit trees were planted to cover the ground areas of the orchard. These measures can help prevent soil erosion, improve micrometeorology of the orchard, and promote fruit production. Besides, rich grass resources here can provide good feeds for the stockbreeding industry.

(3) The virtuous cycle model of fruit–*Amaranthus*–pig–biogas–fruit

This is a new stereoscopic compound model for orchard management initially established in Ningdu County of Jiangxi in 1996. It functions as follows: vacant space between the rows of fruit trees in the orchard is used to plant grain amaranth, which is then used as green fodder for livestock; the breeding industry is thereby developed in the orchards. The manure of animals is put into biogas digesters to produce biogas, providing energy for the economy. The biogas manure (fermented fertilizer) is used to fertilize fruit trees and economic crops in the orchards – a virtuous ecological cycle is formed.

Grain amaranth is a kind of annual xeric crop belonging to Amaranthaceae of *Amaranthus*, with a fast breeding, short growth cycle, high yield, strong adaptability, high nutritional value (the protein content of its stem and leaf exceeds 20%), strong germination ability, a wide range of uses, and other good characteristics. It can be planted or seeded in rows in unoccupied places of the orchard. After a month, its fresh stems and leaves can be removed to feed livestock. The manure from animals can be used to produce biogas, and the biogas fluid and residue can be used as fertilizer for fruit trees and grain amaranth. This model not only reduces the cost of breeding, increases management benefits, and promotes a benign circulation but also allows biological energy to be converted to domestic use.

Being one of the main models for comprehensive management of soil and water conservation, this model utilizes the circulation function of biological chains, forming a new stereoscopic layer, which contributes to the construction of the multi-storey space niche of the orchard, and thus promotes the integration of agriculture, forestry, and animal husbandry and a virtuous circulation of pig–biogas–fruit. This model is suitable for orchards with mature soil and with trees that are beginning to bear fruit and that have relatively strict requirements for water and fertilizer. The model is suitable for forming the basis of a cultivation industry.

10.6.2 Forest–Herb Management Model

10.6.2.1 Site Features

This kind of woodland, located in the low-elevation hilly regions with good site features, belongs to the young timber-oriented forests. Fertile soil, suitable micrometeorology, and water and light conditions here are conducive to the growth of trees and vegetation.

10.6.2.2 Guiding Ideology for Management

By making use of good ecological conditions beneath trees and the open spaces among young and mid-aged trees, one can plant medicinal plants in line with the requirements of the development of the local economy to form the forest–herb compound management model. One can create artificial conditions for the cultivation of natural herbs to make up for the lack of wild medicinal resources. This strategy not only preserves water and soil and improves the ecological environment but also improves the quality of medicine, achieving sustainable development.

10.6.2.3 Key Techniques

(1) Choosing herb species for interplanting
Choose and interplant beautyberry (*Callicarpa bodinieri*), grass coral (*Sarcandra glabra*), and other shade-tolerant herbs or Yellow gardenia, dyers woad (*Clerodendrum cyrtophyllum*), and other species with high economic efficiency demanded by the local market.
(2) Appropriate canopy density of the forests
Illumination level in the forest directly influences the production of medicine. Appropriate canopy density is a prerequisite for the good growth of herbs.

Research shows that for relatively good output, the canopy density for shade-tolerant herbs should be between 0.6 and 0.8. Improper canopy density is not conducive to the growth of herbs. For light-tolerant herbs, we could only interplant them in the young forests before the closure of canopy.

(3) Reasonable planting density of herbs
The planting density should be determined according to the conditions and biological characteristics of particular sites. In areas with good site conditions and careful management, plant herbs in either low density or high density.

(4) Planting appropriate herbs in the right conditions
Many herbs can only grow under special conditions. Therefore, the slope position, altitude, and site conditions should be in accord with the biological characteristics of intercropped herbs and the humidity and illumination conditions required.

10.6.2.4 Application of Models

(1) Interplanting beautyberry among young Chinese firs (*Cunninghamia lanceolata*)

Beautyberry is a kind of neutral to feminine Chinese herbal medicine with unique medical effects and is a type of small perennial shallow-rooted shrub. Since 1991, studies have been conducted on the planting of beautyberry in the newly built Chinese fir forests combined with the afforestation in barren hills.

Various experiments have been carried out, such as tests of planting densities in 3-year-old Chinese fir forests, a comparison test on plantations at different altitudes, and an experiment on plantations with different canopy densities. The results of these experiments show that the yield of beautyberry is closely related to the planting density and the best planting density is 30 cm × 30 cm per plant. Too low or too high a density would each result in low yield per unit area (individuals cannot fully grow if the density is too high). Altitudes also significantly influence the yield of beautyberry. The highest yield is gained in places with an elevation between 350 and 410 m, and yield decreases in accord with the decreases of elevation. This is consistent with the fact that beautyberry requires damp and shady conditions. In addition, the canopy density of Chinese fir forest also has a visible influence on the output of beautyberry and the canopy density of the upper forest is one of the dominant factors that exert an impact on the growth and development of beautyberry. An experiment was arranged to test the growing conditions of beautyberry in five different canopy densities. According to this test, the yield of beautyberry was relatively high when the canopy density was between 0.5 and 0.7 and the most suitable canopy density for planting was 0.6.

It is worth noting that there is a significant correlation between herbs and trees in the forest–herb management system. Beautyberry can effectively contribute to the growth of the Chinese fir trees. The average height of 3-year-old trees was 3.28 m, and the average diameter at breast height was 4.42 cm, significantly different from those growing in pure Chinese fir forests of the same age. At the same time, because beautyberry is a perennial plant with high medicinal value and in urgent market demand, this model can have a great social effect in expanding the production of the *beautyberry* by making full use of forestlands.

(2) Interplanting grass coral in mature Chinese fir forests

Belonging to *Sarcandra* in the Chloranthaceae family, grass coral is a kind of generalized Chinese medicinal herb with extensive use, high benefits, and substantial market demand. Grass coral can be planted in the vacant spaces of the woodland to improve forest productivity, according to its growth and development characteristics and the conditions of woodland resources in the mountainous region of the middle and lower reaches of the Yangtze River. For example, grass coral was interplanted among Chinese fir trees in Lishan woodland in Xingan of Jiangxi Province, which not only made full use of the woodland but also largely maintained natural environmental conditions for growth of grass coral, and thus improved the quality of the herb.

The growth of grass coral requires moderately shady conditions. Therefore, a certain degree of canopy density is needed; however, canopy density also exerts a strong negative influence on the growth of grass coral. Research shows that the best canopy density of Chinese fir forest is 0.7 (Photo 10.14). Canopy density that was too high or too low was not conducive to the growth and quality of grass coral.

Slope positions also influence the growth of grass coral. In order to test whether slope positions exert a conspicuous influence on the growth of grass coral, we furrow cropped grass coral on middle and upper slopes along the contour lines. Ten sample points were chosen at random to measure the average height and the diameter of grass coral. The results of variance analysis showed that significant differences exist in the growth of grass coral planted on different slope positions and that grass coral planted on the mid-slope region grew better than that planted on the upslope.

Photo 10.14 Planting grass coral (a Chinese medicine herb) under Chinese fir forest

10.6.3 Tree–Tea (Camellia sinensis) Compound Model

10.6.3.1 Site Features

Generally speaking, these sites are located in low-elevation hilly regions with relatively good site conditions, higher fertility, and thicker soil layer, and thus have good growth of tea; however, the strong light and low air humidity of pure tea gardens always results in low yield and inferior quality of tea.

10.6.3.2 Guiding Ideology for Management

The guiding ideology of tree–tea compound management is to establish high-yield and high-quality tea plantations either by planting trees in pure tea gardens or by interplanting tea in a sparse woods consisting of tree species. On the one hand, micrometeorological conditions with more diffuse light and higher air humidity can be created with the shade of trees; on the other hand, land resources can also be fully used.

10.6.3.3 Key Techniques

(1) The choice of intercropping tree species

In the tree–tea compound management system, the spatial and temporal locations and functional features of various populations greatly determine whether the relationships between populations are coordinated or not. Therefore, the spatial structure of the intercropping tree garden is very important. Judging from the aboveground level, choose tall and sunshine-loving trees to intercrop with short and shade-resistant tea trees in order to form a rational horizontal mosaic and a vertically layered structure. Judging from the belowground level, the roots of the tall and tea trees should be distributed in different soil layers to form a rationally layered and mosaic structure in the belowground part of the tea garden.

Therefore, the choice of arboreal species determines whether the distribution of spatial locations of the populations is reasonable and tea trees can make full use of light, soil, and other resources. It is the key for the success of this model.

(2) Planting density

The species selected in tree–tea compound management model have partial temporal overlap. This kind of overlap has hardly any effect on tea growth in spring; however, in summer the shade formed by the crowns of tall deciduous trees can help improve the growth and quality of tea. Therefore, a reasonable allocation of temporal scheduling in the tree–tea management model is another key technique for the success of this model.

10.6.3.4 Application of Models

The tea tree, during its long-term evolution, has formed the habit of shade tolerance and prefers high air humidity and more diffuse sunshine. However, in many tea-producing areas, only small areas of suitable terrain are selected to develop tea plantations, due to the lack elsewhere of light and humidity conditions required by high-quality tea. In recent years, experiments have been carried out in the hilly areas of the middle and lower reaches of the Yangtze River to study tea–tree management models, such as the persimmon (*Diospyros kaki*)–tea intercropping model, the slash pine–tea interplanting model (Photo 10.15), the Chinese tallow tree (*S. sebiferum*)–tea intercropping model, Chinese chestnut (*Castanea mollissima*) (Photo 10.16)–tea, the *Eucommia* bark–tea intercropping model among bamboo forests (Photo 10.17), and the wingceltis (*Pteroceltis tatarinowii*)–tea model (Photo 10.18). Good effects have been achieved in all these models, providing feasible technologies for the development of tea production in hilly areas.

Photo 10.15 Pine–tea interplanting model in Huangshan City, Anhui Province, China

Photo 10.16 Chinese chestnut–tea model in Jing prefecture, Anhui Province, China

Photo 10.17 *Eucommia* bark–tea in Tongshan forestry farm, Nanjing, China

Photo 10.18 Wingceltis–tea model in Jing prefecture, Anhui Province, China

(1) Persimmon–tea stereoscopic management model

This trial site is located in Jing County, Anhui Province. These tea plantations were established in 1982 and persimmons (DaFang persimmon) were planted in 1987 by grafting. All the tea plantations had basically the same site conditions and received identical management measures.

Research shows that in the persimmon–tea compound management system, the difference in height between persimmon and tea trees results in a prominent double-layer structure, making tree crowns in a cross and partially overlapping state. Because persimmons are light-loving while tea trees are shade-tolerant, this kind of configuration is good for the full utilization of space and light in the compound plantation. In addition, more than 80% of tea roots are distributed in the soil layer above 20 cm deep, while the feeder–root systems of persimmons mainly concentrate in the soil layer between 20 and 60 cm deep, especially between 20 and 40 cm.

Judging from root breadth and intensity scope, the horizontal distributions of tea and persimmon roots are in a cross and mosaic state. The underground structure is basically reasonable and the utilization degree of soil space in this model is higher than that of the pure tea garden.

During the spring tea period, persimmon is in its leaf unfolding stage and has small shading effects. Therefore, the temperature of the picking surface in a compound management tea plantation is only 0.62°C lower than that of pure tea plantation. During the summer tea period, a significant cooling phenomenon takes place in the intercropping tea garden, with the average temperature of picking surface being 1.95°C lower than that of a pure tea garden. In addition, the daily temperature difference is not as obvious as that of pure tea and the time of high temperature is significantly shortened, which is favorable for the growth of tea shoots. The relative humidity in compound management tea plantations in spring and summer is higher than that of pure tea gardens by 6.2 and 11.3%, which is favorable for the improvement of tea tenderness.

According to biochemical analysis and sensory measurement, the content of amino acids, soluble sugar, and caffeine in tea growing in compound managing tea plantations is higher than that in pure tea plantations, while the content of flavonoids and other compounds which damage the quality of green tea is lower than that in pure tea plantations. In addition, tea growing in compound managed tea plantations is also better than that in pure tea plantations in terms of appearance, smell, taste, and shape.

Moreover, because persimmon trees greatly contribute to the volume of litterfall in the woodland, the content of organic matter and total and available nitrogen in each layer of soil in compound management plantations is higher than that in pure tea plantations. The content of phosphorus and potassium in most soil layers is also higher than that in pure plantations. All these facts indicate that this model improves soil fertility and thus is favorable for the growth of both tea and persimmon.

The persimmon–tea stereoscopic management model largely improves the ecological environment of tea gardens. The relationship between persimmon and tea populations is basically coordinated and the utilization rates of light and soil in this model are higher than those in pure tea models. These show that this model has a relatively good ecological basis. Besides, it also has some other advantages, such as high biomass yield, good quality of tea, and good economic and social benefits.

(2) Pine–tea interplanting model

The trial site is located in BoCun woodland in Huangshan City, southern Anhui Province. The topography of this site is gently sloping with an elevation of 150 m. It was wholly cultivated in 1986 to plant slash pine and loblolly pine with a planting spacing of 6 m × 6 m. Six years later, in 1992 it was cultivated again and two to three rows of tea seedlings were interplanted between every two rows of pines with an individual spacing of 0.3 m.

Research shows that in pine–tea compound management tea plantations, pine trees occupy the upper layer of the space aboveground and intercept the direct component of the solar radiation, leaving mainly diffuse light to shine on the tea crowns. This kind of light has higher long-wavelength ratio and weaker light intensity. Such changes in the forest greatly meet the requirements of tea trees regarding light. If pine trees are in the appropriate density, the relationship of species populations aboveground in the compound management tea garden will be coordinated and the sunshine resource will be fully used. In addition, the feeder–root system of slash pines and loblolly pines mainly concentrates in soil layer between 20 and 60 cm in depth and cross and stratify with the feeder–root systems of tea trees to a certain extent. Although some of slash pine's feeder–root systems are also distributed between the ground surface and 20 cm in depth, and thus compete to some extent with tea trees for water and fertilizer, the overall relationship between two kinds of roots is coordinated and water and fertilizer in all soil layers are fully used.

Because of the shading effect of pine trees, a good micrometeorology is formed in the intercropping garden with suitable temperature, relatively high humidity, and weak light, which provides an ideal ecological environment for the growth of tea and the improvement of tea quality. The relation between tea yield and shading degree can be demonstrated as follows: canopy shading degree 30–40>20–25 (pure tea garden)>0>45–50, which shows that the suitable shading degree is between 30 and 40% in the pine–tea intercropping tea garden and is less favorable for the three other treatments. Severe shading (shading degree is over 45%) will decrease photosynthesis. Therefore, a suitable intermediate density of pine trees is necessary for intercropping tea gardens.

(3) The stereoscopic management model of Chinese tallow–tea

The trial site was located in the Xitian tree farm, Xiuning County of Anhui Province in eastern China, belonging to the low mountainous and hilly region. Chinese tallow was scattered throughout the tea garden. Experiments showed that Chinese tallow was a kind of light-oriented, deep-rooted, and large deciduous tree, while tea was an evergreen, shade-tolerant, shallow-rooted small shrub. In view of their ecological habits and biological characteristics, the composite tea garden of Chinese tallow–tea had a good ecological basis: reasonable allocation of space, harmonious interspecific relationship, and full use of environmental resources. As mentioned above, Chinese tallow and tea were partially overlapping in time but spatially complementary and coordinated with each other. Despite partial overlap in time, there was no fierce competition between the species. As for function, the two made the tea garden ecosystem colorful and meanwhile made the ecological environment conducive to the growth of tea and the improvement of tea quality. It was obvious that in the Chinese tallow–tea composite management system, there existed good niche relationship between the populations; therefore, it was a quite reasonable management model.

(4) The stereoscopic management model of chestnut–tea

The trial site was located in Yansi forest seed farm, Huangshan City. The chestnuts were 11 years old. The density of the seedling afforestation was 390 individual plants/hm^2. Tea plants were interplanted when the chestnuts were 7 years old, point planting with a density of 6,855 individual plants/hm^2. At the same time, a pure tea garden was created in the adjacent area with similar site conditions, row planting with a density of 8,893 individual plants/hm^2. The production, management measures, and tea varieties of the compound management tea garden and pure tea garden were the same.

Test results showed that every year, chestnuts germinated and grew leaves in middle and late March; leaves gradually flourished, starting in mid-April, and began to fall in early-October; then chestnuts turned leafless after mid-November in winter. Chestnut and tea were partially overlapping in time, which was exactly what the tea needed. This model made more efficient use of space and time resources than did pure tea garden and pure chestnut forest, respectively, with a more reasonable time structure. The stereoscopic management model formed by interplanting of chestnut and tea boasted the features of staggering in time, and complex and perfect complementarity in space; the entire tea garden ecosystem made more efficient use of time and space resources than did the pure tea garden. Therefore, it is one of those forest–tea management models that is worth popularizing.

In a word, in the composite management models of persimmon–tea, Chinese tallow–tea, pine–tea, chestnut–tea etc., there exist good niche relationships between populations – complementarily coordinated in space and partially overlapping in time with no fierce competition. As for function, these models made the tea garden ecosystem rich and colorful; they fully improved tea garden ecological environment, making it conducive to tea growth and tea quality enhancement. So they were effective methods and models for establishing high-yield and quality tea gardens.

10.6.4 The Forest–Amaranth–Stockbreeding Composite Management Model with Grain Amaranth as the Linkage

10.6.4.1 Principles of Planting Grain Amaranth in Forestland

(1) The drought tolerance of grain amaranth is the prerequisite that makes its planting in the forestland in question possible. These forestlands and orchards are typically relatively drylands that are usually not irrigated. This makes it impossible in those places for forest (fruit)–agriculture composite management of many excellent crops that are not drought tolerant. Grain amaranth possesses impressive drought tolerance ability. It can withstand extremely dry conditions with water content of 4–6% in the 0–10-cm soil layer. Testing results have shown that the water demand of grain amaranth in the whole growth period was only 41.8–46.8% of wheat, 79.1% of summer cotton, and 51.4–61.7% of summer corn. The rainfall in hilly regions of the middle and lower reaches of the Yangtze River can fully ensure the necessary water for its normal growth. Meanwhile, the vigorous growth period of grain amaranth occurs during the period of ample precipitation in southern China.

(2) With its good characteristics for use of space and time, grain amaranth is a fine crop for interplanting in forestland. Its growth period is short, only 90–120 days.

It can be picked up as green fodder about a month after sowing; it can be cradled 60 days after sowing, with very strong germination capacity after cradling. Therefore, water, soil, and light resources, as well as high temperature, can be fully utilized in forest (fruit) clearance land. In addition, forest/fruit trees' growth needs can be adjusted through the cutting of grain amaranth. The right weeding, cultivation, and fertilization of grain amaranth can adequately provide the needs of the forest/fruit trees. In this way, forest (fruit)–grain amaranth can form a mutually beneficial symbiosis. Besides, grain amaranth can be sown in spring, summer, or autumn with high flexibility; it is convenient for farmers and fruit growers to handle the relationship between production management of forest/fruit trees and a diversified economy.

(3) With widespread use, grain amaranth can help achieve a diversified economy of forestland. Grain amaranth in the seedling stage is a kind of grain with high nutrition. Its straw and leaves during the growing period have good succulence with high protein and can be used to feed pigs, cattle, sheep, rabbits, chickens, ducks, and geese. During the growth period it is also a high-quality green manure. Rich in lysine and essential to humans, the seeds of grain amaranth are an ideal nutritional and healthy food. The straw can be shattered and mixed with corn and early rice to be processed into granular feedstuff, resolving the problem of succulence shortage in winter. These utilizations are effective ways in which forest farms and orchards can develop livestock breeding and processing industries, base long-term benefits on short-term ones, help to realize an organic combination of long-, middle-, and short-term benefits, and promote sustainable and stable development of forest farms and orchards. Grain amaranth also has extensive use in the project of returning farmland to forest/grass in "Western Development of China."

(4) As an important ecological niche resource, grain amaranth is conducive to eco-agricultural engineering construction. Grain amaranth is a high-quality animal feedstuff. One can plant grain amaranth in unoccupied places in forestland (orchards), use it to feed livestock and poultry, and return manure to forestland, orchards, or biogas digesters to produce biogas. The biogas residue and liquid can be returned to forestland and orchards so as to fill up the empty niche in the grass-grazing and detritus food chains, thus making the eco-agricultural engineering of forest/fruit–biogas–pig (livestock and poultry) more stable and efficient.

10.6.4.2 Guiding Ideology for Management

As a high-quality protein feed resource, grain amaranth is introduced to forestland and orchards as a connecting linkage between forestry, agriculture, and stockbreeding, creating a cyclic model among forestland (orchard), grain amaranth, breeding industry, biogas, fertilizer, and forest/fruit tree, in which materials and energy can be used in multiple levels. That is, one can first interplant grain amaranth in the clearance of young forest and orchard; second, grain amaranth can be used as succulence to feed livestock and poultry (pigs, chickens, rabbits, ducks, sheep, fish, etc.) and develop a breeding industry; third, one can return the manure of livestock and poultry to biogas digesters to produce biogas, providing domestic energy; then the biogas (fermented fertilizer) can be returned to forestland and orchards as fertilizer. In this way, a virtuous cycle of the stable composite ecosystem of forestry, agriculture, and stockbreeding can be formed to foster forest growth and breeding industry development, improve the micrometeorology of forestland, make use of soil potential, and achieve the goals of high efficiency, less pollution, low energy consumption, high utilization rate, and sustainable development.

10.6.4.3 Key Techniques

(1) Interplanting
Forestland interplanting of grain amaranth can be achieved by direct seeding or by transplanting seedlings. According to requirements, the planting should be offset from the vertical projection crown of the trees, with appropriate row spacing of 30–40 cm and plant spacing of 20–30 cm.

(2) Cutting
Generally speaking, 2 months after seeding, or 40–50 days after transplanting, when the plant grows to 1–1.5 m height, grain amaranth can be cut as green fodder. But planting and cutting in forestland and orchards should be based on the growth of forest and fruit trees so as not to affect their light needs; in addition, the stubble height should be kept appropriate at about 30–40 cm (too low height results in non-germinating and even to the rotting of stands; too great height leads to low productivity and waste) with 3–5 retaining buds. Fifteen to 20 days later, the second cutting can take place at a slightly higher cutting position than the preceding. Generally speaking, it can be cut 3–4 times, or cut only once with a low stubble of 10 cm, stored green or dried, and processed into pellet feed for reserve.

(3) Utilization
Grain amaranth has many uses, but forestland interplanting mainly involves the following uses:

As vegetables in the seedling period: With seedling height at 10–15 cm, it is edible as a vegetable. Particularly in early spring, it should be covered with plastic sheeting. In southern and middle Jiangxi Province, it is sown from late February to early March. In northern Jiangxi Province, it can also be sown in mid-March, appearing in market earlier than farther south.

Feed as green fodder: Fresh leaves and stems of grain amaranth are quality green fodder for young livestock and poultry; they can be fed directly to animals with appropriate amounts in accordance with their different categories of needs. Because grain amaranth has a unique odor, some animals need an adaptation process.

Feed as dry stem and leaf powder: To solve the problems of green fodder surplus during the vigorous growth period, and the shortage of later straw and feed in winter, the leaves and stems of grain amaranth can be dried, crushed or beaten, mixed with corn and rice flour to be processed into pellet feed, and stored for reserve.

Silage: Grain amaranth can also be stored as silage in its vigorous growth period. But in June and July, there is plenty of rain in southern China, with high humidity and high temperature; anaerobic conditions are hard to guarantee, making silage difficult.

10.6.4.4 Application of the Model

Since 1991, grain amaranth has been interplanted between tree rows in 1-year-old Chinese fir forestland (Photo 10.19), 3-year-old Chinaberry tree forestland, and 2-year-old *Pittosporum tobira*, red maple, red leaf cherry plum (*Prunus ceraifera*), and forestland of sweet olive (*Osmanthus fragrans*) (Photo 10.20). The results showed that due to different light conditions of different stands, the growth environment for grain amaranth varied and so did its growth rhythm. The growth status of grain amaranth interplanted in different young forests is listed in Table 10.1.

Table 10.1 shows that the average height and the crown width of grain amaranth in the Chinese fir forestland exceeded those in the Chinaberry tree forestland, which was related to the canopy characteristics of young needle–broadleaf forests and the light requirements of grain amaranth. The Chinese fir forest was only 1-year old, so the trees did not compete with grain amaranth for sunlight. When the hot part of the season came, the grain amaranth had grown tall; its lush branches would provide shade – the suitable environmental conditions for young Chinese fir. Investigation of the biomass of trees and grain amaranth also showed that the areas of interplanted

Photo 10.19 Interplanting grain amaranth among tree rows in 1-year-old Chinese fir forestland

Photo 10.20 Interplanting grain amaranth in the forestland of sweet olive

grain amaranth had significant differences with those of the control area; the average height of young Chinese fir (30.14 cm) was 14.36 cm higher than that in the control area (15.78 cm). Statistical tests showed that the difference was significant at the 0.01 level. All these indicated a mutually beneficial and complementary relationship between grain amaranth and young Chinese fir forest.

The model not only could increase the utilization efficiency of illumination of forest communities, but also could improve the micrometeorology of forestland and avoid the adverse effects on young Chinese fir of high temperature in summer. The leaves of 3-year-old Chinaberry tree forest were sparse during the budding period, so it exerted little influence on the growth of grain amaranth; when Chinaberry tree grew into its flourishing stage with luxuriant foliage and large canopy density, it provided shade for grain amaranth, limiting its growth and seed setting. In late July, the maximum daily height increase of grain amaranth (type K_{112}) in forestland reached 3.42 cm and that of grain amaranth (type R_{104}) reached 3.90 cm in mid-July. But in this period, grain amaranth in Chinese fir forest stopped growing in height and stem diameter due to high temperature and drought. Although the complementary relationship between Chinaberry tree and grain amaranth was not as good as that between Chinese fir and grain amaranth, it still achieved high biomass. The interplanting in broadleaf forests, if special attention is paid to the adjustment and allocation of tree species, age, plant spacing, and row spacing, will bring substantial economic benefits in the future, either by producing green fodder in early and mid-growth periods of grain amaranth or through interplanting of vegetables or storing green manure in forestland.

Table 10.1 Growth status of grain amaranth interplanted in different young forests (cm)

Tree species	K112				R104			
	Average height	Maximum individual height	Average crown width	Maximum crown width	Average height	Maximum individual height	Average crown width	Maximum crown width
Young Chinaberry tree	100.6	149	47.0	73	90.0	115	50.1	60
Young Chinese fir	106.2	162	56.8	82	120.9	179	59.0	95

10.6.5 The Composite Management Model of Forest–Agriculture (Amaranth)–Stockbreeding in Limestone Mountainous Regions

The natural conditions vary greatly between upper and lower parts of limestone mountains, and the depths of the soil layers are different. Therefore, the productivity of forest soil can only be fully utilized under comprehensive consideration of stereoscopic development, with proper tree species for proper times and places. On the other hand, limestone soil is very fertile, so it has great potential for successful implementation of the forestry-agriculture composite model in the vacant spaces when trees are young.

(1) The cyclic model of forest–amaranth–stockbreeding–forest

The experimental field was located in the limestone mountainous region of Xiushui County, Jiangxi Province. The American grain amaranth was interplanted in the inner parts and on the margins and vacant spaces of the forestland of 3-year-old red stem dogwood (*Cornus stolonifera*), to provide a large quantity of high-quality green fodder for livestock and poultry farming. At the same time, goats were raised in the forest to make use of the favorable conditions in the limestone hillsides: high soil fertility, quick vegetation restoration after forestation, flourishing shrubs, and grass. In addition, amaranth was sown in the undergrowth and picked up during different periods or cut many times, by which fresh stems were obtained and used to feed pigs, goats, fish, etc., partly replacing concentrated feed. Then, large quantities of manure and feces from the breeding industry were returned to the forestland, promoting the growth of trees.

(2) The cyclic model of fruit–amaranth–fish–fruit

Nursery and economic fruits can be developed in the foothills; amaranth can be interplanted inside the forestland; ponds can be excavated to raise fish with amaranth; the mud of ponds is returned to orchards; all these together form a virtuous cycle. According to the analysis of the stereoscopic management in limestone regions of Xiushui County, the cyclic system composed of different models with long-, medium-, and short-term benefits is reasonable, scientific, and feasible (Fig. 10.5). As a project with short-term benefits, the forestry–agriculture management can achieve good economic benefits before the maturation of trees; meanwhile it can also improve soil quality. As projects with medium-term benefits, the forest–medicine model and economic forest have a long harvest time and good economic returns, and play an important role in capital turnover. As a main provider of long-term benefits (ecological and economic), the tree–shrub composite model plays a significant role in improving the harsh habitat condition of limestone mountains; meanwhile, it can allow the cultivation of a large quantity of commodity timbers, which confer long-term benefits. In addition, the changeable colors, shapes, and types of broadleaf trees, integrated with rare rock and karst cave landscape on the mountains, boast high ornamental value; thus extra income can be made by opening up forest parks. The virtuous ecological cycle of forest–amaranth–stockbreeding model in Fig. 10.3 combined agriculture, forestry, and stockbreeding of limestone

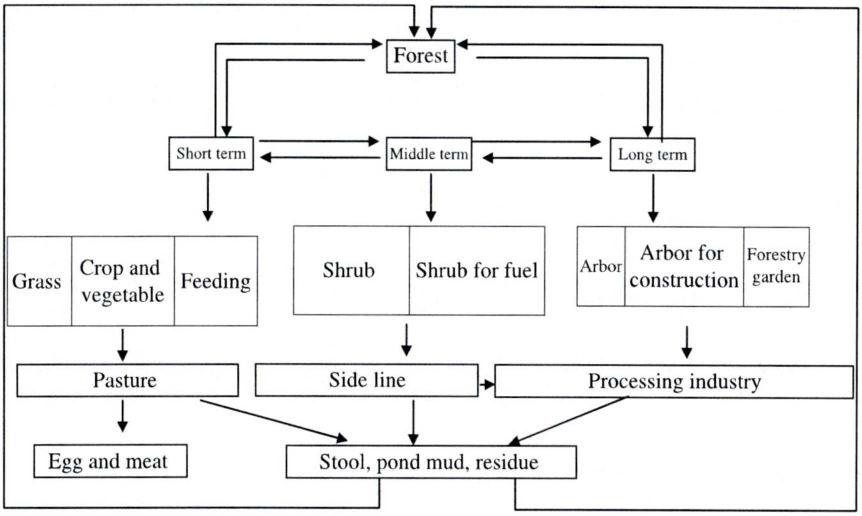

Fig. 10.5 Analysis of the composite management model of an ecological demonstration forest farm in Xiushui County

mountains into a whole, forming a small material cyclic system, fully utilizing available time and space, and achieving multi-level material utilization inside the system.

Figure 10.5 shows that these models were independent, though closely related to each other; they are systems on their own, while at the same time parts of the composite ecological system of multi-level, multi-function, multi-benefit, and stereoscopic management on limestone hillsides. It is the organic combination of long-, medium-, and short-term projects that integrate the economic, social, and ecological benefits of forestry together and form a sequence of planting, breeding, processing, and management, thus achieving the high productivity of stereoscopic forestry and ensuring the successful composite development of limestone mountains.

(3) The fruit–amaranth and the stockbreeding–biogas–fruit management models

1. Analysis of the models
   The introduction of amaranth made the material and energy flow of the system more reasonable, increased the biological chain links, and optimized the structure, forming multi-level utilization of materials inside the system. In this way, it made full use of the orchards' light energy and nutrients, provided plenty of high-quality fodder for stockbreeding, and achieved the objectives of replacing cultivation with cropping, improving soil quality, preventing soil and water loss, and improving micrometeorology, thus promoting the growth of fruit trees. Meanwhile, it could also utilize waste, reduce pollution, and provide energy and good fertilizers. What was especially important was that amaranth could be used as a substitute for some concentrated feeds, reducing farming costs, plus being simple to plant and low in cost. It also increased organic fertilizers necessary to fruit trees, promoted the development of breeding industry, and brought

about the flourishing of the fruit industry. All these advantages made it possible for the system to achieve a maximum in biomass production, part of which was processed by animals and returned to forests. This increased the intermediate links of the food chain, improving material utilization and conversion rates and providing more products for the market. Then the capital was reinvested in forests, orchards, agriculture, and other plantings to acquire more products and reap greater benefits (Fig. 10.6).

2. Application of the models
   The models were implemented in a navel orange orchard in Ningdu County, Jiangxi Province. Starting in early spring, amaranth was interplanted between the fruit trees' rows in different phases, and meanwhile the breeding industry was developed – raising pigs, chickens, rabbits, fish, and other livestock. After growing for about 40 days, amaranth was cut every day for 3 months to feed pigs, chickens, rabbits, fish, and other livestock. The manure of livestock and parts of the amaranth straw were collected in biogas digesters to produce biogas, providing farmers with energy for domestic purposes like illumination. The biogas slurry was used to spray fruit trees, feed pigs, and water amaranth. The biogas residue was returned to orchards to increase fertility, promoting the growth of fruit trees and improving soil quality. Due to the introduction of grain amaranth (an excellent fodder with rich protein) into the virtuous cycle of the ecological chain of agriculture and stockbreeding, edible meat, eggs, and milk were produced by secondary producers – that is, by poultry and domestic animals feeding on forage, straw, and tail materials. The manure of poultry and livestock

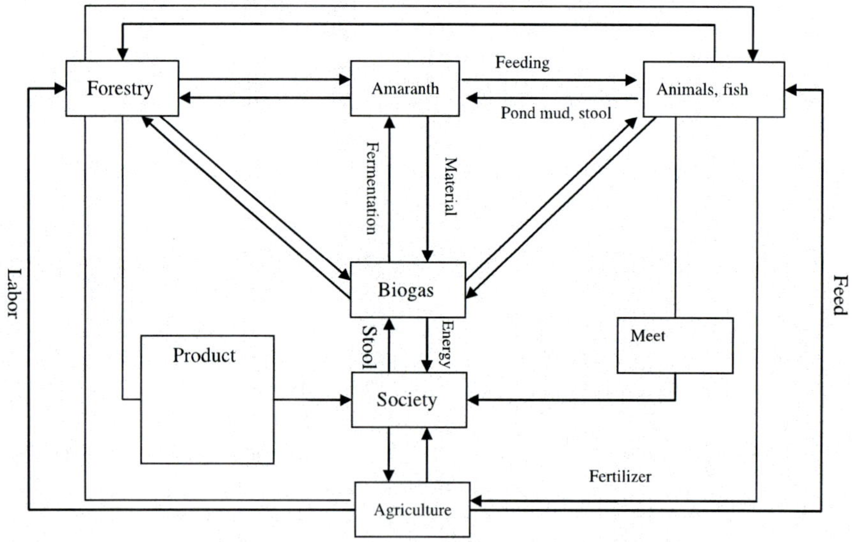

Fig. 10.6 Cyclic model of forest–amaranth–stockbreeding and fruit–amaranth–stock

produced gas and fertilizers in the conversion and transmission processes of the detritus food chain. In this way, it achieved a material cycle and high utilization rate of nutrients in the whole agricultural ecosystem. Therefore, grain amaranth played an important ecological niche role in grass-grazing food chain and detritus food chain, acted as a link integrating forestry–agriculture–stockbreeding, filled the ecological niche gap, and improved the relationship between basic nutrition and secondary production, thus making up the existing deficiencies in the current state of the fruit industry.

Now the biogas project, which provides not only gas energy for cooking and lightening but also organic fertilizer for crop, is widely carried out in the countryside throughout China. For each new biogas tank, the owner can get subsides of 800 yuan (which is enough to make a biogas tank in China) from the government. Photo 10.21 shows a peasant in the Shangshe village is standing beside his newly made biogas tank to take a rest near his new house (left in Photo 10.21) and old house (right in Photo 10.21) (2009, 8–22). Before his biogas tank was made, straw and animal stool from his family was formerly piled up beside (Photo 10.22) the road and caused pollution and eutrophication of water. And he also went up the mountain to cut twigs and collect pine litter for use as fuel for cooking. But with the biogas project and solar water heater project in the Chinese countryside, he said he would feel ashamed to cut twigs and collect pine litter as fuel again, which is thought to be a hard and inefficient work. Second, the biogas project uses straw and animal stool as material so that animal stool was no longer piled up everywhere in the village, making the environment in the countryside clean. Furthermore, straw firing, which polluted air and caused forest fire in the past, has decreased greatly.

Photo 10.21 A peasant in the Shangshe village is standing beside his newly made biogas tank taking a rest near his new house (2009, 8–22)

Photo 10.22 Straw and animal stool were usually piled up beside the countryside roads before the biogas tank was used

10.7 Management Models of Commercial Forests with Good Site Features

10.7.1 Management Model of Commercial Forests

10.7.1.1 Site Features

The sites under consideration encompassed areas with a thick soil layer (above 60 cm) and a thick humus layer (above 20 cm), boasted good physical and chemical soil properties, and were located on intermediate to downhill slopes, with a site index above 14. Thanks to good hydrothermal conditions of such sites, forests grew rapidly and could be used for commercial products.

10.7.1.2 Guiding Ideology for Management

Forest category (timber forest, industrial material forest, or economic forest) and proper tree species were determined based on the following principles: the full utilization of superior site conditions of such forestlands, comprehensive consideration of different site conditions, different management objectives, and management practices based on ecological–economic principles and requirements of regional economic development. Intensive management was adopted to pursue the goals of rapid growth, high yield, and high efficiency.

The main objectives of vegetation restoration and management in such forestlands were to artificially promote progressive succession, rapidly restore the forest community, and create timber forest, industrial material forest, and economic forest. Its guiding ideology for management was first to design good species configurations to maintain the stability of the ecosystem, based on the full exploration of suitable local plant species, and second to develop an optimal scheme and implementation measures according to the features of different tree species and stand structures. Special attention should be paid to intensive management and application of advanced science and technology, especially the optimal configuration of the composite structure of trees, shrubs, and grasses, and the creation of mixed forest, with an aim of building reasonable complementary relationships among species, fully utilizing the regional hydrothermal resources, restoring forests, promoting the growth of trees, and achieving the objectives of good quality, high efficiency, and rapid growth.

10.7.1.3 Key Techniques

(1) Strict site control and genetic control
When selecting tree species, the focus should be on superior provenance to adopt excellent species.
(2) Emphasis on seedling and planting quality
Choose appropriate soil preparation measures according to soil, vegetation, climate, and other factors, and apply fertilizers reasonably and in a timely manner to guarantee the survival rate of the forest and promote the growth of trees.
(3) Introduction of associated tree species (especially nitrogen-fixing trees)
Retain the original vegetation belt, adopt the mixed model of coniferous and broadleaf trees, to enrich the composition and structure of vegetation types, maintain biological diversity, enhance the resistance of the system, and protect and enhance land capability of forests, achieving the goal of sustainable management.
(4) Reasonable structure regulation
Reduce the initial planting density, postpone the thinning age, increase the thinning efficiency, improve stumpage specification and quality, reduce management costs, and increase management benefits. Improve the thinning, final cutting operations, cleanup methods in forestland, and increase the restitution of nutrients and accelerate nutrient circulation.
(5) Maintain land capability; combine management and development
Introduce plant species with economic value or soil-improving capability; construct the forestry–agriculture–stockbreeding composite model with stereoscopic management of forest, fruit, and medicine, integrate long-, medium-, and short-term benefits, and balance the ecological and economic benefits.

10.7.1.4 Application of Models

(1) High-efficiency intensive culture of industrial timber forest

The national forestry programs by the state are called "Expansion of forestry in eastern China and soil erosion control in western China" and "Let the land take a rest in northern China and utilization of forest in southern china." The hilly regions in the middle and lower reaches of the Yangtze River assume responsibilities of both "eastern expansion" and "southern utilization." In view of the distribution of site conditions of existing forestland, it is appropriate to create fast-growing and high-yielding forest in hilly scrub, as well as in deteriorated and regenerated forestlands. Therefore, taking the slash pine and loblolly pine in Anhui Province as an example, a study was made to explore the fast-growing and high-yielding technologies in southern forests, to investigate the management models of commercial forest in hills and mountains, and to propose a series of matching technologies for intensive cultivation. This included such steps as choosing proper trees for certain sites, adopting high-quality seeds, cultivating sound seedlings (cultivation of both "bare-root" sound seedlings and container sound seedlings), carefully managing soil preparation and construction of forest tracks and fire protection belts, introducing associated tree species, applying basic fertilizers, reinforcing seedling cultivation, topdressing young forest, controlling pests, and cultivating and thinning in a timely manner.

The introduction of associated tree species to create a mixed forest is the key to the success of high-efficiency intensive culture of industrial timber forest in the hilly regions of middle and lower reaches of the Yangtze River. The study of high-efficiency intensive culturing of slash pine and loblolly pine industrial timber forests indicated that the introduction of associated tree species was conducive to the stability of ecosystem. On the one hand, it alleviated many ecological problems such as soil degradation, pest and disease damage, forest fires, and soil and water loss, caused by the uniform structure of large areas of artificial industrial forests. On the other hand, by retaining part of the original vegetation and cultivating certain amounts of associated tree species, it enriched the composition and structure of vegetation types, maintained biological diversity, enhanced system resistance, and preserved and enhanced land capability, thus achieving the goal of sustainable management. Following the three basic principles of adaptability, ecological economy, and multi-purpose of forest, red *S. superba*, Chinese tulip tree, *Sassafras* tree, southern wild jujube, sawtooth oak (*Quercus acutissima*), golden larch, Chinese honey locust fruit (*Gleditsia sinensis*), and fine black locust, which has good nitrogen-fixing capability, were selected as associated tree species for the establishment of an artificial industrial forest. In addition, the configuration of associated species promoted the mixing patterns of belts and blocks instead of among individual plants or between the lines of trees. Meanwhile the management strictly followed the principle of proper trees for certain sites.

Another important aspect is to build forest trails and fire barrier zones. Forest trails should be built based on topographical features of forestland, and these also serve as fire barrier zones. The main forest trail should be set along the main ridge, with two or three rows of fire-resistant trees such as *S. superba* planted on both sides; a branch trail should be set along the branch ridge, with one or two rows of *S. superba*, etc. planted on both sides as a branch fire barrier zone. When preparing soil for the establishment of forest, a certain percentage of the original vegetation belt at the top, slope, foot, and other key parts of mountain should be retained, with a vegetation bandwidth at the top of 10–20 m, while the bandwidths at the slope and foot of mountain are usually set according to slope length (generally 2–5 m). As for natural shrubs and herbaceous plants in artificial forests, as long as they do not hinder the growth of planted trees, they should also be nurtured with care.

(2) The mixed model of coniferous and deciduous trees in low hilly regions

Straight in stem, good in quality, and high in production rate, Chinese fir, masson pine, and exotic pines (slash pine, pitch pine (*Pinus rigida*), loblolly pine, etc.) are the main species for forest establishment in the hilly regions of middle and lower reaches of the Yangtze River. The large-area engineering afforestation also adopts Chinese fir and pure pine forests as its principal management types. But the litter of coniferous trees is high in lipids and acid materials, which easily forms crude ground cover and humus, deteriorates soil physical and chemical properties, and causes soil degradation. Because there is a lack of natural enemies of pests, both diseases and pests occur frequently in pure coniferous forest and the stands are unstable, so it is not conducive to the sustainable and stable development of forestry. With a large quantity of leaves and ash content, broadleaf trees can improve soil physical and chemical properties, and maintain and enhance soil capability. Experimental fields of different site types were chosen in Jiangxi Province for the study of mixed models of coniferous and deciduous trees based on the principles of ecological and economic coordination.

Mixed combinations: Based on different site conditions and management objectives, different broadleaf species and mixing modes were chosen to establish different mixed forests of coniferous and deciduous trees. Guided by the main purposes of restoring vegetation, conserving water and soil, and maintaining land capability, sweet gum and *Lespedeza* were usually selected; in order to gain higher economic benefits and more timber, tung oil tree and *S. superba* were then selected. The actual combinations included the following: slash pine × sweet gum (Photo 10.23), slash pine × tung oil tree, Chinese fir × tung oil tree, slash pine × *S. superba*, slash pine × *Lespedeza*, masson pine × oak (Photo 10.24), etc.

10.7.1.5 Effects of the Models on Yield and Growth

(1) The mixed model of pine–tung oil tree

Tung oil tree is an important industrial woody oil tree species, with favorable features of fast growth, early fruit bearing, and high yield. The mixed forest of

Photo 10.23 Slash pine × sweet gum in Anhui Province, China

Photo 10.24 Masson pine × oak in Anhui Province, China

coniferous and deciduous trees, with tung oil tree as the broadleaf species, not only improves soil quality, maintains land capability, and achieves better ecological benefits but also produces better short-term economic benefits, and combines long-term benefits with short-term benefits, thus achieving the organic integration of long- and short-term benefits. The Tung oil trees in the mixed model in Yiyang experimental field began to bear fruits 3 years after afforestation, with the maximum fruits per individual tree exceeding 20 and the average yield being 244.5 kg ha^{-1} and then 652.5 kg hm^{-2} in the fourth year, increasing over 166.9% from the previous year. Currently tung oil is in short supply at the market, with a price of over 20 yuan per kg. If harvested for 3–5 years before the major species reach the fast-growing period, 1950–3250 kg of tung oil can be harvested per hectare (assuming the average annual harvest is 650 kg hm^{-2}), acquiring a short-term revenue of 2,600–4,333 yuan hm^{-2}, that is to say, the economic income during harvesting time can reach 865.5 yuan per hectare annually. Considerable quantities of wood are available after the trees are cut down. Another fast-rewarding and highly beneficial multi-management method is to use the wood to cultivate mushrooms.

(2) The mixed model of pine–*S. superba*

The *S. superba* is a main species in forests in southern China. With a straight stem and hard woody texture, it serves as a special material for the textile industry and timber for commercial furniture. Mixed with conifers trees, it can be used as a fire barrier zone due to its thick, leather-like, and fire-proof leaves, playing a role in preventing forest fire and pine caterpillar damage in pine forest and generating great economic benefits both directly and indirectly.

(3) The mixed model of pine–sweet gum

Sweet gum is a fast-growing deciduous timber species, boasting the features of extensive adaptability and drought resistance. Due to the fine wood structure, pest-resistant feature, persistence in shape, and smooth surface, it is the timber species for plywood, food boxes, tea boxes, and paper making. As a pioneer species for creating mixed forest of coniferous and deciduous trees in regions with poor site conditions, it can effectively increase soil organic matter, improve soil quality, maintain land capability, and effectively curb the spread of caterpillars. In addition, it has high ornamental values with red and yellow leaves in autumn.

(4) The mixed model of pine–*Lespedeza*

The *Lespedeza* is a multipurpose tree species. With rapid growth, suckering flexibility, and high yield, it is a good fuelwood species. With luxuriant foliage and plenty of litter, it provides protection for the soil surface, preserves soil moisture, and improves the cycling of soil nutrients; therefore, it is a good species for soil and water conservation. With a large quantity of nitrogen-fixing rhizobia and high content of ash in litter, it is a good species for improving soil quality and fertility. Therefore, in seriously eroded areas that lack fuelwood, it can be used for mixed forest development, achieving good ecological benefits and short-term economic benefits.

Results showed that the establishment of mixed forest with *Lespedeza* and sweet gum in seriously eroded areas with poor site conditions could effectively increase vegetation coverage and under-story biomass. This mixed forest helped to conserve soil and water, increase soil organic matter, improve soil physical and chemical properties, and maintain soil fertility, contributing to the sustainable, stable growth and high yield of forests, and achieving good ecological benefits.

The reasonable selection of broadleaf species should be based on the full use of growth differences among species and nutrients in different soil layers and on the improvement of the utilization rate and productivity of forestland. Experiments showed that root systems in the models of pine–sweet gum, pine–*S. superba,* and pine–tung oil tree were coordinated and complementary. Among these, pine and tung oil trees had a relatively poor complementary relationship, but this did not have a significant negative influence on the growth of young trees.

As for forestlands with good site conditions, the fast-rewarding broadleaf trees (e.g., tung oil tree) with high economic benefits should be selected in the establishment of forests. The establishment of mixed forests with coniferous and deciduous trees can achieve satisfactory short- and medium-term economic benefits, help to support long-term benefits with short-term ones, and properly combine short-, medium-, and long-term benefits together, thus achieving synchronous growth of ecological and economic benefits.

10.7.2 High-Efficiency Intensive Culture of Dual-Purpose Bamboo Forest with Shoot and Timber Orientation

Bamboo is an important forest resource in hilly regions of the middle and lower reaches of the Yangtze River and occupies a very important position in China's forest production. In recent years, due to the soaring demand and increasing chopping, nutrient depletion in bamboo forests has increased steadily, and land capability has been severely decreasing, which has led to the low productivity of bamboo forest and has seriously threatened the sustainable management and utilization of bamboo forest.

10.7.2.1 Key Techniques

Taking soil fertility maintenance in nutrient management of bamboo forest as the main objective, simple and applicable diagnostic methods were adopted to determine the constraints of soil nutrients in bamboo forest; intensive management and nutrient management measures were established based on soil fertility, the pattern of requirement of fertilizer by bamboos, and different management goals. Using the optimum balanced fertilization formula and technology suitable to the management of bamboo forest in Jiangxi Province, fertilizers were applied at proper times and in proper quantities according to the critical periods at which bamboos require fertilizer, the variability of soil nutrients, structural changes of the stands, and the nutritional status of bamboo. Dynamic nutrient management, precise fertilization, and nutrient information management with computers were adopted to meet the requirements of optimal growth of bamboo, as well as economic growth, in order to realize the efficient sustainable management of bamboo forest.

The model mainly included such steps as the restructuring of bamboo forest, mixed management of bamboo with broadleaf trees, intensive cultivation and management, nutrient regulation in major or minor years (in China, the year for fruit trees yielding much more fruit than the average yield is called the major year, that much more smaller than average yield is called minor year, and that near the average is called average year), and maintenance of soil fertility in forestland. The concrete factors were as follows: patterns of fertilizer requirement in major or minor or average years and of bamboo forests with different purposes (shoot oriented, timber oriented, or shoot-and-timber oriented), the identification of constraints on soil nutrients in bamboo forest, optimum fertilization formula for different site conditions and different management models (major year, minor year, and average year), the relationship between bamboo age and growth of bamboo forest, corresponding technologies and management models for intensive management of bamboo forest, diagnostic technologies for nutrition of bamboo leaves, and nutrient management technology with computers.

10.7.2.2 Application of the Model

The model can be applied to bamboo forest in southern China, especially for the intensive cultivation and sustainable high-yielding management of timber-oriented and shoot-oriented bamboo forests (Photo 10.25). For the design of temporal sequences, the unification of restructuring and functional benefits of bamboo forest should be emphasized, in order to improve the distribution of bamboo at different ages and increase average annual output. The results showed that currently in about 15% of the bamboo forest in southern China with intensive management, balanced fertilization and high-efficiency intensive cultivation were the important means to achieve goals of high, stable yield, and good quality. After balanced fertilization, the production of bamboo forest and the diameter at chest height of new bamboo increased markedly. The diameter increased by 11.2% annually; the income per kilogram of fertilizer increased by 11.8 yuan. The net income increased by 5,804

Photo 10.25 Timber-oriented and shoot-oriented bamboo forests with intensive management in Anhui Province

yuan per hectare, with the input–output ratio at 1:5, indicating quite remarkable economic benefits. Through the intensive quantitative nutrient management, nutrients in bamboo forest were balanced, achieving the biggest economic benefits with the most reasonable and scientific management technologies.

*10.7.3 High-Efficiency Intensive Culture of High-Quality Oil Tea (*Camellia oleifera*) Forest*

Oil tea is a unique subtropical oil tree of China. The current cultivation area is over 4 million ha, distributed over 500 counties and cities of 17 provinces. With an annual oil production more than 120 million kg, it is a very important resource for economic forests and the preferred species for the comprehensive development of southern mountainous and hilly regions. The middle and lower reaches of the Yangtze River are the main producing areas of oil tea. At present, there exist widespread extensive management with low production and unscientific fertilization in large areas of oil tea forest in south China; the production potential of oil tea is far from being fully developed.

10.7.3.1 Key Techniques

Key techniques in this model include the following: selection of good clone species of oil tea, intensive management of fertilizers, pruning technologies, nutrient management technologies (balanced fertilization), flower and fruit management technologies, pest and disease control technologies, and composite management technologies.

10.7.3.2 Application of the Model

Experimental fields were located in the main production areas of Jiangxi Province, namely Yichun, Suichuan, Jingdezhen, Shangrao, Yingtan, and Gi'an. Based on the study, a series of corresponding technologies and models of intensive cultivation were proposed for different management levels and types, such as culturing techniques of excellent clones, transformation of low-yield forest, and renewal of aged forest. The study showed that red soil in southern China generally lacked P and K nutrients; therefore, the nitrogen-fixing, P- and K-increasing fertilization formulae and composite stereoscopic management model with intensive cultivation measures should be adopted (Photo 10.26).

Meanwhile, according to the constraints of soil nutrients in oil tea forestland of different management models, the relationship between oil tea production and the availability of soil nutrients should be made clear, thus working out reasonable fertilization quantities, methods, time, frequencies, etc. Special emphasis should be put on the composite forestry–agriculture management during the early growth period and the flower, fruit, and nutrient management during the full fruit period, in order to achieve intensive nutritional management of the ecosystem, reasonably determine nutrient management measures, and select appropriate management models.

In this way, the quality and productivity of oil tea forest can be enhanced; soil quality can be restored and improved; soil deterioration and the current discrepancy between supply and demand of oil tea can be solved, achieving sustainable development of oil tea forest, resulting in maximum economic benefits with minimal costs, and helping farmers to shake off poverty and set out on a road to prosperity.

Photo 10.26 Intensive cultivation measures by intercropping oil tea with sesame

Chapter 11
Effect of Afforestation on Soil and Water Conservation

Abstract Reforestation of bare land has a good effect upon our environment. It has usually been assumed that reforestation can control erosion. However, in the early years of reforestation, things were complicated. Experiments showed that after the tending of young trees, runoff clearly increased until 2–3 storm events had occurred. Soil loss increased within 1 month after staddle tending and it was especially large during storm events. It was estimated that if there are 250–1500 fish scale pits per hectare for tree planting on sloping land, then 362.5–725 m^3 hm^{-2} $year^{-1}$ sediments could be conserved in place. The amounts of nutrient elements in the surface runoff were highest in the bare land; the amounts were second highest in Chinese fir (*Cunninghamia lanceolata*) forest, and those in the level terrace areas were the lowest, indicating that changes in the topography and improvements in vegetation coverage had an important effect on nutrient loss. Second, reforestation can improve biodiversity. The understory plant, physical, chemical, and microbial properties were investigated for several types of forests over 10 years of reforestation at Yiyang County and Taihe County. Before reforestation, there were only shade-intolerant herb species. After reforestation, some shade-loving plants that like a wet, shady environment were found growing well. Third, reforestation has a good effect on micrometeorology. After reforestation the maximum temperature inside the forest in summer was 2.3°C lower than that outside the forest, and the relative humidity increased by 9%. The daily variation amplitude of temperature inside the forest was 2.8°C lower than that outside the forest; the amplitude of variation of humidity increased by 5.5%.

11.1 The Amount of Soil Erosion in Different Types of Lands

In Chinese fir (*Cunninghamia lanceolata*) woodland areas, terraced fields, and hilly bare land, three plots for the observation of surface runoff and soil loss, the basic conditions of which are listed in Table 11.1, were set up in the Xiashu ecological site in southern Jiangsu Province. For the USLE plot, a collecting tank of 1.5 m ×1.2 m ×1.2 m and a collecting tank of 1 m ×1 m ×1 m were built. Sediment and surface runoff enter the first collecting tank, and then enter the second one when the first tank is full. In front of the first tank, a sharp-crested weir of 3° was set up. An auto-graphic rain gauge was set up at each weir to measure the surface runoff.

J.C. Zhang et al., *Theory and Practice of Soil Loss Control in Eastern China*,
DOI 10.1007/978-1-4419-9679-4_11, © Springer Science+Business Media, LLC 2011

Table 11.1 Basic condition of plots in the Xiashu forestry farm

Site	Plant coverage	Slope	Slope direction	Soil types	Area (m^2)
Chinese fir	0.8	15°	E	Yellow-brown soil	150
Level terrace	Newly planted tea garden, interplanting soybean	Level terrace	W	Yellow-brown soil	2000
Bare land	Uncovered	12°	SW	Yellow-brown soil	100

Table 11.2 Water and soil loss at each plot

Monitoring site	Annual average of runoff depth (mm)	Runoff coefficient (%)	Annual soil loss [t/(km^2 a)]	Comparison to that of bare land (%)
Chinese fir forest	162.75	15.5	281	134.01
Level terrace	213.15	20.3	480	82.25
Bare land	604.4	48	16414.5	3340.8

To a large extent, topographical condition and vegetation cover condition influence the amount of surface runoff and soil erosion. Table 11.2 lists the annual water and soil loss of each plot.

Table 11.2 shows that both the runoff coefficient and the annual soil loss on bare land without control measures were clearly higher than those of level terrace and Chinese fir forest. The surface runoff coefficient on bare land was 48%, which was 2.36 and 3.1 times higher than those level terrace and Chinese fir forest. Soil erosion losses were, 16,414.5 t/(km^2 a), 34.2 and 58.4 times of those of level terrace and Chinese fir forest, respectively. It can be concluded that terrace and forest can greatly decrease soil erosion.

At the same time, research on soil erosion of several forest types has been carried out in hills in southern Jiangsu Province. According to the synchronous measurements of soil erosion on each plot, annual soil loss is 14414.5 t km^{-2} in bare land, 281 t km^{-2} in Chinese fir forest, 255.4 t km^{-2} in loblolly pine (*Pinus taeda*) forest, and 238.4 t km^{-2} in oak forest. In Chinese fir forest, soil erosion was 1.95% of that in bare land, in loblolly pine, it was 1.77%, and in ring-cupped oak (*Cyclobalanopsis myrsinaefolia)* forest, it was 1.65%.

In China, values of 200, 500, and 1000 t·(km^{-2} a^{-1}) were set as the permitted soil losses in northeastern north China, the hills of southern China, and the non-obvious erosion areas of Loess Plateau, respectively. Using the measured data, average annual soil losses were estimated. Predicted values of soil loss in the Chinese fir forest, the loblolly pine forest, and the oak forest were found to be within the limits of permitted soil loss. Soil erosion on bare land was much higher than the values of permitted soil loss, since the extent of soil erosion was severe. Soil loss control through reforestation was very effective.

In order to compare the soil erosion of different years after reforestation, a plot method was used to test soil erosion in 6 years in newly planted slash pine (*Pinus elliottii)* forest (Table 11.3) at Matou forest farm in Anhui Province. The ground surface had been severely destroyed, resulting from stub digging at *clear-cut* in the first year of reforestation. In a control area, only the mowing of grass and chopping down of shrubs was carried out. This latter "almost no disturbance" plot produced 27.6% more runoff than did the reforestation plot. Annual soil loss was 195.59 t km^{-2} in the reforestation plot and 284.35 t km^{-2} in the control plot. The former can be classified as a non-obvious erosion area, while the latter belongs to the category of a light erosion area. The amount of water and soil loss clearly decreased in the following 5 years, and annual soil losses in the reforestation plot were slightly more than those in the control plot in the second, third, and fourth years, because of tending young stand, but the values still fell within the range of non-obvious erosion. From the fifth year, there was no obvious difference between the reforestation plot and the control plot.

At the same time, effects of tending on water and soil loss in reforestation land of slash pine were measured and the results are shown in Table 11.4. Young stand tending was carried out on 30th May 1993. Soil loss events 1 month before and after 30th May 1993 showed that after tending, runoff clearly increased and then after 2–3 storm events, runoff in the two plots tended to be equal. Soil losses increased within 1 month of staddle tending and it was especially large during storm events. As time went by, soil losses decreased (Fig. 11.1).

Nine plots had been selected as monitoring sites for soil loss observation, located in Shadi in Gan County, Guifeng in Yiyang, Xiushui, Taihe, and five other places. The experiment, which started in 1991, indicated that almost all types of reforestations on barren land could greatly control soil loss.

According to the observation data in December 1997, in the newly planted slash pine plantation, the average height was 3.1 m, the DBH (diameter at breast height, 1.3 m) 5.5 cm, and the crown width 1.7 m × 1.8 m. The understory plants were bamboo, *Smilax china*, *Rhododendron*, etc. The coverage was 60% and the average height was 1.0 m. Besides the slash pine plot, another plot with similar plants, but without slash pine, with a coverage above 90% was selected as monitoring site, where the average height of plants was 1.5 m. With the growing of trees and plants, the runoff decreased gradually, and soil loss clearly decreased as well (Table 11.5).

Based on field observation data in the Ganzhou monitoring site, a combination of engineering measures were used, including physical measures, such as level-ditch preparation and fish scale pits, and biological measures, such as choosing pioneer species with high adaptability, for example, masson pine (*Pinus massoniana*), herbs and shrubs, *Amorpha* (*Amorpha fruticosa*), *Lespedeza* (*Lespedeza bicolor*), and Bahia grass (*Paspalum notatum*). Then trees, shrubs, and grass were planted together in a reasonable combination. These measures were effective in providing multiple layers for the interception of rain and for decreasing both surface runoff and soil erosion. Runoff accumulated in fish scale pits and ditches, which could improve the water conditions in forestland. This method also facilitated the survival of forest and understory growth of vegetation.

Table 11.3 Annual variation of soil and water loss of slash pine

Year	Times of runoff	Rainfall (mm)	Runoff		Soil loss		Runoff coefficient	
			ZL ($t\ hm^{-2}$)	CK ($t\ hm^{-2}$)	ZL ($kg\ hm^{-2}$)	CK ($kg\ hm^{-2}$)	ZL (%)	CK (%)
First year after plantation	38	645.30	442.996	610.794	1955.8632	2843.5567	6.86	9.47
The second year	48	1136.20	363.001	447.647	552.8566	489.0866	3.19	3.94
The third year	26	595.90	178.157	167.369	147.8482	72.0540	2.99	2.81
The fourth year	39	1040.20	240.777	184.364	50.2091	27.0632	2.31	1.77
The fifth year	30	1038.20	343.020	354.951	71.8652	68.1221	3.30	3.42
The sixth year	31	650.50	43.320	47.578	5.8355	5.5183	0.67	0.73
Total	212	5105.30	1611.241	1812.683	2784.4777	2502.8807	3.16	3.55

ZL means the slash pine forest and CK means the bare land

Table 11.4 Water and soil loss before and after tending young stand

Date		Precipitation (mm)	WZL (t hm^{-2})	CK (t hm^{-2})	SZL (kg hm^{-2})	SCK (kg hm^{-2})	WZL/WCK	SZL/SCK
BS		25.30	3.700	3.800	3.700	3.640	0.97	1.01
	1993-05-06	28.40	4.500	4.500	4.500	4.50	1.00	1.00
	1993-05-09	14.10	1.300	1.315	1.0400	0.9731	0.909	1.07
	1993-05-14	13.90	0.280	0.285	0.1120	0.1111	0.98	1.01
	1993-05-16	8.70	0.090	0.099	0.0450	0.0505	0.91	0.89
	1993-05-19	70.80	48.000	48.100	48.0000	44.7330	1.00	1.07
	1993-05-20	36.10	4.800	4.800	5.2800	5.0880	1.00	1.04
	1993-05-28	10.20	0.230	0.238	0.1150	0.1190	0.97	0.97
	Total	207.50	629.00	63.137	62.7920	59.227	1.00	1.06
AS	1993-06-01	35.00	14.100	9.300	28.9050	10.8810	1.52	2.66
	1993-06-2	31.70	12.900	8.800	24.5100	8.7120	1.47	2.81
	1993-06-13	8.90	0.090	0.094	0.0603	0.0404	0.96	1.49
	1993-06-15	23.10	3.900	4.100	9.0870	5.0020	0.95	1.82
	1993-06-18	20.00	2.700	3.500	4.7790	2.8000	0.77	1.71
	1993-06-21	11.40	0.685	0.815	0.5823	0.4482	0.84	1,30
	1993-06-22	39.00	30.800	29.300	65.9120	26.9560	1.05	2.45
	1993-06-30	28.60	9.800	10.300	15.4840	9.0640	0.95	1.71
	Total	197.70	74.957	66.209	149.3196	63.9037	1.13	2.34

WZL and WCK are runoff in slash pine plot and control plot, respectively; SZL and SCK are soil loss in slash pine plot and control plot, respectively; AS, after tending young stand; BS, before tending young stand

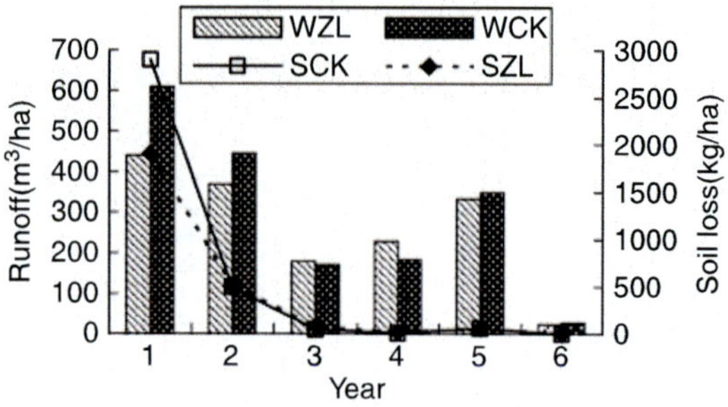

Fig. 11.1 Variation of annual soil and water loss

In the Gan prefecture, Shadi monitoring site, soil losses that accumulated in fish scale-shaped pits (both the diameter and the depth were 120 cm), located in the upper, middle, and lower slopes, were investigated and are shown in Table 11.6.

It was inferred that if there are 250–1500 fish scale pits for one hectare, then 362.5–725 m^3 hm^{-2} sediments could be conserved in place.

11.2 Loss of Soil Nutrient Elements in Different Types of Reforestations

After reforestation, the quantity and the quality of nutrient elements in the rainfall are changed to a large extent by leaching of the canopy and the removal of soil nutrient by surface runoff. From analysis of surface runoff quality, physical, and chemical properties of sediment at monitoring plots in the main reforestation hills at the middle and lower reaches of the Yangtze River, the concentrations of N, P, and K of the surface runoff, as well as soil organic matter content, were obtained for single rainfall events (Table 11.7). These data were used to calculate the amount of nutrient element losses by single rainfall events and the annual soil loss. It is shown in Table 11.7 that the nutrient concentration in surface runoff of each area was higher than that of natural precipitation, except for the element nitrogen (N). Compared with forestland, the concentrations of N and P in runoff in bare land were higher than those from forestland and clear-cut of oak forest, whereas the concentrations of K, Ca, and Mg were lower than those from forestland. This was because the element N is difficult to carry in dissolved form through organic matter. When rain falls from the forest canopy to the ground, water is absorbed by each layer and N is taken up by the litter, which makes the N concentration of runoff much lower than that in precipitation. In the bare land, N concentration of runoff was a little higher than that in precipitation because runoff water dissolved N. The content of P in the natural precipitation was very low, but that in surface runoff was higher because of the effect of leaching from each layer of forest. The element K is easily dissolved by precipitation, so the content of K in the forest was higher than that of clear-cut of oak forestland, bare land, and natural precipitation. The content of K of oak forest was the highest. Ca and Mg are also easily dissolved by runoff,

Table 11.5 General conditions of plots in Jiangxi Province

Monitoring site	Guifeng in Yiyang	Shadi in Gan County	Taihe	Xiushui	Jingdezhen	Yujiang
Geomorphology	Hill	Low hill	Hillock of low hill	Low hill	Low hill	Hillock of low hill
Slope (°)	2–10	15–30	<10	15–35	10–25	5–350
Altitude (m)	36–55	120–320	74–125	<321	30–100	86–202
Soil	Red soil	Red soil	Red soil	Neutral yellow, red soil, and red lime soil	Red soil and yellow, red soil	Red soil
Soil parent materials	Red sandstone	Granite	Killas, shale, and sandstone	Limestone	Phyllite, granite	Shale, sandstone
Soil thickness	Thin	Thin	Unequal	30–100 cm	60–100 cm	Thin
Fertility	Extremely barren	Extremely barren	Barren	Medium	Medium	Extremely barren
Gravel content.	Little	Many	More	Gravel content 30–50%; bareness degree of rock 59–70%	30–40%	20–80%
Vegetation condition	Almost no vegetation	Average coverage 10%	No arbor and bush, but few uneven grass	Few arbor in barestone gaps, bush and grass with block distribution	Good vegetation such as masson pine, Chinese fir	Almost no vegetation on slope top, coverage of inferior slope 30%
Erosion condition	Red desert	Proliferated erosion ditch and collapsing hill	Severe soil erosion on partial area	Severe	Mild	Severe

Table 11.6 Soil loss in fish scale-shaped pit at Shadi in Gan County (m^3)

Sample number	1	2	3	Average
The upper slope	0.6	0.55	0.62	0.59
The middle part of slope	0.33	0.75	0.15	0.53
The foot of slope	0.55	0.23	0.20	0.59

Table 11.7 Average nutrient concentration in surface runoff

	Average nutrient concentration (mg kg^{-1})					
Element	Rainfall	Bare land	Chinese fir forest	Loblolly pine	Deciduous oak forest	Clear-cut of oak forest
N	1.150	1.266	0.611	0.651	0.774	0.842
PO_4^{3-}	0.045	0.572	0.275	0.090	0.448	0.498
K^+	1.290	2.856	5.400	6.030	10.566	4.925
Ca^{2+}	3.750	2.000	4.650	3.200	2.200	3.533
Mg^{2+}	0.872	0.400	0.937	2.282	1.832	1.552

Table 11.8 Annual surface runoff and nutrient element export in each monitoring site

		Nutrient elements (kg hm^{-2} a^{-1})				
Vegetation types	Surface runoff [m^3 (km^{-2} a)]	N	P	K	Ca	Mg
Bare land	931.94	1179.8	533.1	2661.6	1863.9	372.8
Chinese fir	196.36	120.0	54.0	1060.3	913.1	184.1
Ratio to that of bare land (%)	21.07	10.2	10.1	39.8	49.0	49.4
Loblolly pine	801.09	521.5	88.1	4830.6	2563.5	1828.4
Ratio to that of bare land (%)	85.96	44.2	16.5	181.5	137.5	490.4
Deciduous oak forest	123.56	95.6	55.4	1305.6	271.8	226.4
Ratio to that of bare land (%)	13.26	8.1	10.4	49.1	14.6	60.7
Clear-cut of oak forest	220.8	185.6	74.4	1087.2	780.0	342.6
Ratio to that of bare land (%)	23.7	15.7	14.0	40.9	41.9	91.9

so the contents of the Ca and Mg in forest sites were higher than those in natural precipitation and bare land.

Water and soil loss through runoff is a very important pathway of element export. Serious water and soil loss usually leads to decline of soil fertility and finally a decrease in the growth rates of plants. Table 11.8 shows the relationship between surface runoff and export of nutrient elements at each monitoring site.

Table 11.9 Relationship between soil erosion and nutrient element export

Vegetation types	Soil erosion kg (hm^{-2} a^{-1})	Soil organic matter (kg hm^{-2} a^{-1})	N (kg hm^{-2} a^{-1})	P (kg hm^{-2} a^{-1})	K (kg hm^{-2} a^{-1})
Bare land	14390.90	266.23	16.26	8.06	171.11
Chinese fir	281.00	9.98	0.45	0.11	3.63
Ratio to that of bare land (%)	1.95	3.72	2.75	1.40	2.12
Loblolly pine	225.4	4.16	0.25	0.13	2.68
Ratio to that of bare land (%)	1.57	1.56	1.57	1.56	1.57
Deciduous oak forest	238.40	7.99	0.44	0.07	2.67
Ratio to that of bare land (%)	1.66	3.00	2.70	0.88	1.56
Clear-cut of oak forest	537.90	18.02	0.98	0.16	6.02
Ratio to that of bare land (%)	3.74	6.77	6.03	1.99	3.52

Table 11.8 shows that the loss of nutrient elements in different forest types varies greatly. The more the surface runoff that was produced, the more the nutrient elements exported. In the loblolly pine forest, runoff was higher than that from the other forest types, and loss of K, Ca, and Mg was higher than that of bare land.

Soil erosion is one of the main causes of land degradation. The bare land not only suffers from raindrop impact, which causes damage of soil texture, but also loses abundant nutrient elements by soil loss. Table 11.9 shows that the loss of nutrient elements increased with the soil erosion amount. The order of soil loss according to amount was the following: bare land > clear-cut of oak forest > Chinese fir forest > oak forest and pine forest.

Table 11.10 shows the nutrient elements in soil loss in Chinese fir forest, tea (*Camellia sinensis*) garden interplanted with soybean (*Glycine max*), and level terraces on hills in southern Jiangsu Province. The amount of nutrient elements in the surface runoff was highest in the bare land. Losses of the nutrients N, P, K, and organic matter in the bare land were the highest among three land use types, corresponding to the loss of sediment. The second highest was that of Chinese fir forest, and the lowest was that of the level terrace. This indicated that changes of topographic conditions and improvement of vegetation coverage had an important effect on nutrient loss. Losses of total nitrogen and phosphorus in level terraces were 34.74 and 29.51% of those in the bare land, respectively, while those in Chinese fir forest were 27.93 and 22.98% of those in the bare land, respectively. In the Chinese fir forest, soil organic matter was abundant due to litter decomposition. Furthermore, leaching of the canopy also led to the loss of organic matter. Losses of available K and organic matter in the Chinese fir forest were 5.68 and 9.3% of those in the bare land, respectively, but were higher than those in the level terrace area, which were 33.6 and 2.45% of those in bare land, respectively. Losses of total N, available P, K, and organic matter in terraces were 67.26, 70.49, 66.4, and 97.57% lower than those in bare land, respectively, while those in Chinese fir forest were 72.07, 77.02, 43.18,

Table 11.10 Comparison of loss of nutrient elements in different types of land use

Project		Surface runoff		Sediment			
		Concentration (%)	Amount (kg)	Concentration (%)	Amount (kg)	Total (kg)	Ratio to that of bare land (%)
Terrace	Total nitrogen	2.064	4399	0.0837	68.84	4467.8	32.74
	Available P	0.435	927	0.0399	8.2	935.2	29.51
	Available K	20.178	1075.2	27.832	57.23	1142.4	33.6
	Organic matter	/	/	1.6871	1387.64	1387.64	2.43
Chinese fir forest	Total nitrogen	2.236	3639	0.01294	172.87	3811.87	27.93
	Available P	0.455	701	0.181	27.408	728.04	22.98
	Available K	26.019	1058.6	25.56	856.32	1914.92	5.68
	Organic matter	/	/	2.9564	650.14	650.14	9.3
Bare land	Total nitrogen	1.869	11296	0.0704	2351.92	13647.92	100
	Available P				1882.29	3169.3	100
	Available K				20886.15	33699.25	100
	Organic matter	/	/		6990.7	6990.7	100

and 90.7% lower than those in bare land. It can be concluded that terrace and forest prevent soil from deterioration, which in turn assures the development of agriculture and forestry.

In order to compare the loss of soil nutrient elements in different years after reforestation, newly planted slash pine forest was selected for the location of monitoring sites (Table 11.11). According to the field observation data of water and soil loss amounts and its nutrients, it was calculated that the amount of loss of organic matter and the main nutrients N, P, and K from a reforestation zone was 131 kg hm^{-2}, while that from bare land was 153 kg hm^{-2} during the observation period. The results in detail are shown in Table 11.11. The main components of the nutrient losses were soil organic matter and total K. Ratios of organic matter and total K to the total amount were 55 and 26%, respectively. The main components of available nutrient losses were available K and available N, most of which were lost in the form of runoff liquid, taking 10 and 5% of the total amounts, respectively. The condition of nutrient loss indicated that this is an area of extreme depletion of phosphate. The rule of nutrient loss was coincident with soil loss.

11.3 The Effect of Reforestation on Plant Biodiversity

In the monitoring sites in Jingde City, management measures such as closing hillsides, were used as part of forest planting. In China, in order to culture forest in hillsides where there are usually some young trees and shrub, people closed the hill and stipulated that no utilization, such as pasture, fetching wood, and wood fuel, was permitted within 5–10 years or longer period. Broadleaf trees such as sweet gum (*Liquidambar formosana*) and red *Schima superba* (*Schima wallichii*) were maintained, while conifer trees, such as slash pine and pond pine (*Pinus rigida*), were planted. With these measures, the native plant types were maintained, which helped to avoid damage to the soil and plants by site preparation measures. At the same time, mixed forest helped to form a multi-layer canopy and a thick litter layer, which can intercept precipitation, decrease surface runoff and soil loss, and conserve water resources. In the monitoring sites in Jingde City, field observation data showed that litter in the closing culture forest was 13.8 times that in the plantation forest and its water conservation capacity was 5.8% higher than that in the slash pine plantation forest.

In a monitoring site in Yujiang prefecture, Jinjiang Town, comprehensive management on the upslope, mid-slope, and downslope areas, with trees, bushes, and herbs in stereocombination, was used to form a forestry–agriculture–pasture recycling ecosystem. That is, on the upslope, pioneer trees, such as pine, were planted to control soil loss, as well as for fuel material; on the middle and downslopes (lower), slash pine and red maple (*Acer rubrum*) were planted; at the foot of the mountain, chestnut (*Castanea mollissima*), peach (*Amygdalus persica*), pear (*Pyrus pyrifolia*), and orange trees (*Citrus reticulata*) were planted. Peanuts (*Arachis hypogaea*), some green folder plants, and rape (*Brassica campestris*) were planted among the fruit

Table 11.11 Annual variation of nutrient loss in newly planted slash pine forest

Annual treatment project		Organic matter	Total N (kg hm^{-2})	Total P (kg hm^{-2})	Total K (kg hm^{-2})	Available N (kg hm^{-2})	Available P (kg hm^{-2})	Available K (kg hm^{-2})
Growth in the first year	ZL	51.349	1.682	0.743	23.901	2.012	0.043	4.065
	CK	69.098	2.104	0.882	33.44	2.966	0.033	3.044
The second year	ZL	14.54	0.475	0.21	6.756	1.546	0.033	3.044
	CK	11.824	0.36	0.151	5.722	2.027	0.028	35.99
The third year	ZL	3.888	0.127	0.056	1.807	0.746	0.016	1.46
	CK	1.751	0.053	0.022	0.847	0.748	0.01	1.933
The fourth year	ZL	1.32	0.043	0.019	0.614	0.995	0.022	1.933
	CK	0.658	0.02	0.008	0.318	0.819	0.011	1.441
The fifth year	ZL	1.89	0.062	0.027	0.878	1.417	0.031	2.753
	CK	1.655	0.005	0.021	0.801	1.578	0.021	2.755
The sixth year	ZL	0.153	0.005	0.002	0.071	0.001	0	0.002
	CK	0.134	0.004	0.002	0.065	0.001	0	0.001
Total	ZL	73.231	2.395	1.058	34.026	6.716	0.145	13.256
	CK	85.03	2.592	1.086	41.194	8.139	0.111	14.588

ZL means the slash pine forest and CK means the bare land

Table 11.12 Growth condition of the main forest stand in an experimental garden

Forest stand	Age (a)	Average height (cm)	Average diameter (cm) at ground	Canopy density	Leaf area index	Monitoring site
Chinese toon	4	511	7.02	Closed	2.127	Xiushui (planting density: 2160 hm^{-2})
Sichuan alder	4	654	7.66	Closed	7.59	
Orientalis	4	233	3.48	0.4		
Red maple	4	323	4.55	0.6	5.604	
Slash pine	4	189	6.30	0.6		Guifeng in Yiyang (planting density: 2505 hm^{-2})
Masson pine	4	151	3.90	0.4		Shadi in Gan County (planting density: 2505 hm^{-2})
Pond pine	4	165	4.75	0.8		Taihe (planting density: 2505 hm^{-2})
Loblolly pine	4	250				Yujiang (planting density: 2505 hm^{-2})

trees. The plants gradually recovered, starting from the foot of the mountain to the top. Before the management measures were taken, soil loss on this slope was very serious. At a dam close to the monitored slope, the water had extremely high turbidity after precipitation. It was estimated from field observation data that annual soil loss was 4 mm. Around the dam, there were 75.2 hm^2 of bare land, which exported more than 4500 t of sediments into the reservoir at the dam. Four years after reforestation, from mountain top to foot, plant coverage reached 63–90%. Soil and water loss decreased greatly after storms, as did runoff to the dam after storms.

Limestone barren hill experimental sites were located in Xiushui prefecture. The bareness of rock ranged between 50 and 70%. So fast-growing trees, for example, *Cedrela sinensis* (*Toona sinensis*), *Alnus cremastogyne* (*Alnus cremastogyne*), *Ailanthus* (*Ailanthus altissima*), Formosan sweet gum (*L. formosana*), dogwood (*Cornus macrophylla*), phoenix tree (*Firmiana simplex*), and Wheel wingnut (*Cyclocarya paliurus*), which are adaptable to limestone hill areas, were selected. Shrubs such as *Amorpha* and *Lespedeza* were also used, together with the trees listed in Table 11.12.

Broadleaf tree species have dense branches and leaves, litter, and root systems, which protect the ground surface from the impact of raindrops. Finally, the function of water conservation in forestland was strengthened and soil loss was controlled.

In experimental sites of Guifeng in Yiyang, Shadi in Gan County, Taihe, and other places with serious soil losses, mixed forests composed of conifer and broadleaf trees were planted. At the same time, shrubs and fodder grass were also planted to construct a stereoscopic combination with trees. That is, conifer trees like masson pine, pond pine, slash pine; shrubs like *A. cremastogyne* and *Lespedeza*;

Table 11.13 Growth of understory vegetation

Monitoring site	Stand structure	Coverage Before greening (%)	After greening (%)	Biomass (dry weight) (t hm^{-2})
Taihe	Slash pine × red maple	30	90	7.99
Guifeng	Slash pine × *Lespedeza*	0–5	56.7	2.45
Shadi in Gan County	Slash pine × *Amorpha* × grass	5–10	85	7.63

and grass like broadleaf paspalum (*Paspalum wettsteinii*), large crabgrass (*Digitaria sanguinalis*), ryegrass (*Lolium perenne*), goose crown grass (*Roegneria kamoji*), and Bahia grass were planted. All these were believed useful to speed up the recovery of plant covering undergrowth and control of water and soil loss. The growth conditions of understory plants in three monitoring sites in Taihe, Guifeng, and Shadi in 1994 are shown in Table 11.13.

Ten years after reforestation, different types of forests were selected in the deteriorating red soil areas located at Yiyang County and Taihe County, for which the understory plant, physical, chemical, and microbial properties were investigated and analyzed. Then the effect of reforestation on water and soil conversation after 10 years was discussed.

During the process of reforestation of forests on degenerating red soil, the understory plant species were increasing and the species composition was changing. The recovered species are shown in Table 11.14 in detail. Before reforestation, there were only intolerant herb species such as orange grass (*Cymbopogon goeringii*), silk grass (*Imperata koenigii*), *Setaria* (*Setaria viridis*), wild old grass (*Arundinella hirta*), yellow grass (*Heteropogon contortus*), and sunny plants or hygromesophytes plants such as Daphne genkwa and dyers woad (*Clerodendrum cyrtophyllum*). After reforestation, some hygromesophytes and shade-loving plants such as moss, fern, and other shrubs or herb species that like a wet, shady environment, such as black pearl vegetable (*Lysimachia heterogenea*), Taiwan glume (*Agrostis sozanensis*), climbing fern spore (*Lygodium japonicum*), mouse grass (*Gnaphalium affine*), creeping *Liriope* (*Liriope spicata*), and bean curd plant (*Premna microphylla*), were found growing well. The local plants like camphor (*Cinnamomum camphora*) and blue oak with small leaves (*Cyclobalanopsis glauca*) were starting to bud too. All these introduced species proved that after restoration for 10 years, the soil water and nutritional conditions were greatly improved, and the effect of water and soil conversation were prominent, which provided suitable environmental condition for local plants' natural regeneration.

The forest ecosystem can reduce surface runoff and prolong the time of release of runoff by multi-layer interception. Consequently it can affect flood detention. The understory litter has the same effect on mitigating runoff, because litter on the forest floor increases the land surface roughness and Manning coefficient. Canopy cover, sapling density, litter depth, and woody debris appear to be important ecological

Table 11.14 Variation of undergrowth species in the process of forest rebuilding

Plant species	1991	1993	2001	Plant species	1991	1993	2001
Lichenes			+	*Rhamnus crenata*			+
Dicranum scoparium			+	*Euscaphis japonica*		+	
Funaria hygrometrica		+		*Toxicodendron succedaneum*			+
Hypnum plumaeforme			+	*Centella asiatica*			+
Pogonatum inflexum			+	*Symplocos groffii*		+	+
Plagiogyria japonica			+	*Serissa serissoides*		+	+
Dicranopteris linearis			+	*Uncaria rhynchophylla*			+
L. japonicum			+	*Lonicera japonica*			+
Stenoloma chusanum			+	*Gnaphalium adnatum*			+
Pteridium aquilinum var. latiusculum		+	+	*Lactuca indica*			+
Adiantum flabellulatum			+	*Erigeron annuus*			+
Woodwardia japonica			+	*Stimpsonia chamaedryides*			+
Dryopteris chinensis			+	*L. heterogenea*			+
Berberis virgetorum		+		*Adenophora paniculata*			+
C. camphora			+	*Monochasma savatieri*		+	
Viola philippica			+	*Scoparia dulcis*			+
Drosera peltata		+	+	*Vitex negundo var. cannabifolia*	+	+	+
Oxalis corniculata		+	+	*P. microphylla*			+
Daphne genkwa			+	*C. cyrtophyllum*			+
Actinostemma tenerum			+	*Caryopteris incana*			+
Camellia oleifera			+	*Schizonepeta tenuifolia*		+	+
Vaccinium bracteatum	+	+	+	*Salvia kiangsiensis*			+
Melastoma dodecandrum			+	*Clinopodium chinensis*			+
Grewia biloba			+	*Hemerocallis fulva*		+	
Glochidion fortunei	+		+	*S. china*	+	+	+
Glochidion daltonii			+	*L. spicata*			+
Rosa laevigata	+	+	+	*S. viridis*	+		+
Rosa cymosa		+	+	*A. hirta*	+	+	+
Rubus parvifolius		+	+	*H. contortus*		+	
Rubus corchorifolius			+	*Imperata cylindrica var. major*		+	
Potentilla discolor			+	*Prunella vulgaris*		+	
Buxus microphylla var. sinica			+	*C. goeringii*			+
C. myrsinaefolia			+	*Themeda triandra*			+
Ficus pandurata			+	*A. sozanensis*			+
Ilex cornuta			+				

"+" indicates the species which has appeared

factors that determine the magnitude of soil loss. Trees provide and maintain a litter layer, which protects the soil against the impact of raindrops. Zhang et al. (1994) reported that the Manning coefficient showed an exponential relationship with litter in plantations of Chinese pine and black locust (*Robinia pseudoacacia*) forest. When litter was 20 t hm^{-2}, the Manning coefficient was 0.1995, 0.1784, 0.1297 in populus forest, locust forest, and oil pine (*Pinus tabuliformis*) forest, respectively. Cheng et al. reported that runoff speed was proportional to slope and amount of runoff, and was in inversely proportional to the amount of litter.

Pure forest of pine of medium density with at least 1.382 t hm^{-2} litter on the ground, or thin pine forest with an undisturbed moss layer on the ground, could each decrease surface runoff.

11.4 Improvement Effect of Reforestation on Micrometeorology

11.4.1 The Improvement Effect of Rehabilitated Forest Ecosystem on Micrometeorology

In order to control soil erosion, recover vegetation and the forest community rapidly, promote the progressive succession of remaining forests, and improve the ecological environment species for the five different types of degraded barren ecosystems in Jiangxi Province, we adopted the principles of adaptation to local conditions and used favored trees for suitable land, comprising long-, medium-, and short-term benefits, and rational configuration of forest categories and trees. A comprehensive control policy was taken: combining trees, shrubs, and grasses; mixing coniferous and broadleaf forests; integrating forestry, agriculture, and husbandry; and paying equal attention to ecological and economic benefits. Appropriate experimental research about restoration and reconstruction were conducted according to different characteristics. The micrometeorological conditions of the interior forest improved significantly (Table 11.15) after 4 years of effective regulation. Table 11.15 indicates that vegetation coverage and undergrowth plant biomass of different types of degraded barren ecosystems increased greatly after the comprehensive treatment; vegetation coverage of severely eroded barren hills in Gan and Guifeng counties was 10 times bigger than that of the original value, and the climatic conditions in forestland had changed greatly. In summer, the maximum temperature inside the forestland was 4–19°C lower than that outside the forestland; the relative humidity increased; the variable amplitude of soil temperature decreased; and forest growth reached or approached a normal level, thus achieving the preliminary goal of restoring and rebuilding the forest ecosystem.

At the same time, according to the types and distribution conditions of red soil erosion on inferior land in hilly regions of Jiangxi Province, typical inferior land originating from four different soil parent materials was chosen in four test sites – Shadi in Gan County, Guifeng in Yiyang County, Jinjiang water reservoir in Yujiang, and Ningzhou in Xiushui County. Guided by the main objective of reconstructing the ecological environment, afforestation experiments on multi-layer structures of trees, shrubs, and grasses under deteriorated environmental conditions, adaptability experiments on different tree and grass species, mixture combination experiments

Table 11.15 Change of micrometeorological factors of the interior forest after restoration and reconstruction

Types of degraded barren hills	Test points	Highest temperature (°C)	Lowest temperature (°C)	Relative humidity (%)	Change range of soil temperature (°C)
Severely eroded	Contrast area	44.0	24.0	76.7	30.5–33.4
	Interior of forest	41.1	23.4	80.5	29.0–30.0
Bare limestone	Contrast area	54.9	23.6	78.3	29.0–30.8
	Interior of forest	35.7	23.1	84.4	25.8–27.6
Gravel and thatch	Contrast area	49.9	39.2	75.5	27.6–34.6
	Interior of forest	44.2	26.4	82.7	26.6–34.3
Hill with flourishing shrub	Contrast area	32.2	22.4	76.2	28.4–32.2
	Interior of forest	32.3	23.2	78	25.9–30.1
Hillock with sparse boscage	Contrast area	51.4	24.0	69.7	30.8–32.4
	Interior of forest	49.5	23.0	71.0	29.9–31.5

on different tree species, application experiments of different afforestation technologies, and experiments of forestry–agriculture composite management models were all carried out. After 4 years of treatment, obvious biological benefits were achieved.

Sampled micrometeorological data before and after afforestation in Table 11.16 showed the following for the four test sites (area of about 300 hm^2): the seriously eroded area has become green in 4 years; plant coverage has increased by 50–60%; and formerly barren hills have been occupied by animals like hares and mountain mice. The temperature inside the forest was 3–6°C lower than that outside the forest; the humidity increased by 10%; and the soil temperature amplitude of variation decreased. All these not only were conducive to the growth of woods and movement of soil microorganisms, eliminating the effect of temperature change on soil erosion, but also adjusted the climate over a large region, achieving the preliminary goal of improving the ecological environment.

In addition, in the experimental area for soil and water conservation located at Guifeng in Yiyang County, comparative observations of micrometeorology in forestland with a tree–shrub–grass stereoscopic combination, and in control areas without forest, were conducted in the fourth year after afforestation (August 16, 1994). Each climate factor was observed every 2 h from 6:00 to 20:00. Air temperature, relative humidity, ground temperature, and maximum and minimum temperatures of the ground surface were all measured. Test results are listed in Table 11.17.

According to Table 11.17, after reforestation the maximum temperature inside the forest in summer was 2.3°C lower than that outside the forest, and the relative humidity increased by 9%. The daily variation amplitude of temperature inside the forest was 2.8°C lower than that outside the forest; the relative of humidity increased by 5.5%; the variation amplitude of soil temperature became conspicuously smaller.

Table 11.16 Statistics measuring the results of environmental factor changes in each site

Test sites		Micrometeorology					
		T_{max} (°C)	T_{min} (°C)	Daily range (°C)	Relative humidity (°C)	Difference value (°C)	Variation amplitude of soil temperature (°C)
Yiyang	Interior of forest	40.2	24.5	15.7	85.8	5.5	29.0–31.0
	Exterior of forest	42.5	24.0	18.5	80.3		30.5–31.6
Gan County	Interior of forest	42.0	22.3	17.7	75.1	2.1	29.3–30.8
	Exterior of forest	45.5	24.0	21.5	73.0		31.3–33.4
Xiushui	Interior of forest	35.7	23.1	12.6	84.4	6.1	25.8–27.6
	Exterior of forest	54.9	23.6	31.3	78.3		29.0–30.8
Yujiang	Interior of forest	42.5	24.0	18.5	83.5	4.4	30.8–32.0
	Exterior of forest	43.0	24.0	19.0	79.1		30.5–32.2

Table 11.17 Statistics of measuring results of ecological factor changes after afforestation at Guifeng

Test sites	Micrometeorology						
	Maximum surface temperature (°C)	Minimum surface temperature (°C)	Daily range (°C)	Difference value (°C)	Relative humidity (%)	Difference value (%)	Variation amplitude of soil temperature (°C)
Interior of forest	40.2	24.5	15.7	2.8	85.7	5.5	29–31
Exterior of forest	42.5	24.0	18.5	–	80.3	–	30.5–51.6

Apparently, the above results were due to the complementarity and shading effects of tree–shrub–grass multi-layer structure, which played a facilitating role in moderating the forestland temperature, avoiding the burning of roots and crowns of trees by high temperature, and reducing the influence of severe cold, heat, dry, and wet conditions of soil on soil erosion, and decreasing the transpiration of trees. Meanwhile, the evident improvement of micrometeorology fully showed that the slash pine × *Lespedeza* × grass multi-layer structure of arbor–shrub–grass was especially conducive to the treatment of "red desert." Merely planting trees or grass failed because the single ecosystem layer could not adapt to the adverse environment; only a multi-layer structure was able to overcome this disadvantage and enable the adverse environment of red desert to change from a simple unharmonious ecosystem to a complex ecosystem with a benign cycle, which, consequently, led to successful treatment. For example, Bermuda grass, ryegrass, and *R. kamoji* were planted several years ago at the test site in Guifeng when the treatment procedure of planting grass before trees was attempted. But most of the grasses died because of the adverse environment; only a few patches remained in scattered places. After the planting of slash pine and *Lespedeza*, the remaining grass quickly recovered growth and propagative ability, especially the Bermuda grass demonstrated a strong growth advantage and soil fixing effect. Thanks to the rapid coverage of forestland by grass, the forestland environment improved and plants began to grow. The number of plant species increased to over 10; birds and mammals also returned. These indicated from one aspect that the multi-layer structure, the treatment manner of combining trees, shrubs, and grasses, and the coordination of tree species were all excellent models for successful soil and water conservation, as well as for improving adverse soil conditions.

11.4.2 The Improvement Effect of the Circulation System of Forestry–Agriculture–Husbandry on Micrometeorology

In 1994, the forest–amaranth–livestock–forest cycle model (Guo and Niu, 2000, 2002) was used for a newly established young forest in the limestone barren hills in Xiushui: an excellent feed, grain amaranth (*Amaranthus hypochondriacus*), was interplanted under 3-year-old trees, while at the same time a livestock breeding industry was developed, raising 18 pigs, 40 sheep, and 6000 fish. We adopted interval sowing, stage picking, and multiple cutting, harvesting large quantities of grain amaranth green stems and leaves to feed pigs, fish, and sheep, and partially replacing concentrated feed. The mud and livestock manures of pigs and sheep were returned to forestland to increase fertility and accelerate growth of trees and amaranth. At the same time, we adopted fruit–amaranth–poultry and livestock–biogas–fruit cycle models at a navel orange orchard in a forest station at Tiantou, Ningdu. The study showed that the forestry–agriculture–husbandry cycle model achieved not only good economic benefits but also satisfactory ecological benefits.

Table 11.18 The changed micrometeorological conditions of the "benign cycle" experimental forest

	C. macrophylla		Navel orange	
Forest stand	Test forest	Control area*	Test forest	Control area*
Organic matter %	3.68	3.39		
Bulk density (g m^{-3})	1.29	1.34	1.28	1.34
Field moisture capacity (%)	31.1	23.8	39.73	34.56
Total porosity (%)	56.8	40.5	50.93	46.34
Air permeability of soil	33.97	26.03	21.58	7.01
Average daily temperature (°C)	29.5	30	29.8	29.9
Relative humidity (%)	73	72	80.7	79.3
Variation amplitude of ground temperature (°C)	27.1–34.8	25.5–34.5	28.7–30.9	29.4–33.0
Lowest surface temperature (°C)	23	25	16.9	16.6
Highest surface temperature (°C)	47.8	50.2	45.5	47.8
Surface temperature (°C)	33.3	34.3	34	33.5
Remark: investigation time	1994-8-17		1996-7-30	

*Remark: control area refers to the forests without interplanting grain amaranth; measurement of micrometeorology used the average value of the results observed every 2 h from 6:00 to 20:00

The ecological benefits of the forestry–agriculture–husbandry favorable cycle was manifested in many aspects, such as comprehensive soil improvement, soil fertilization, water and soil conservation, and improvement of forestland and orchard ecological environment by various factors of the cycle. Experimental results (Table 11.18) for the "favorable cycle" system of forestland and orchard, and the corresponding control areas, showed that the implementation of the forest (fruit)–amaranth–livestock–biogas–forest (fruit) cycle model could effectively improve the ecological environment of forestland and orchards. Meanwhile, the average temperature inside the forest in summer decreased; the relative humidity increased; the variations in amplitude of air and ground temperature became smaller; and the highest temperature in summer was 2.3°C lower than that of control area. All these created a favorable ecological environment for the growth of forest trees and fruit trees, and exerted a positive influence on the micrometeorology around this area.

The Forestry College in Jiangxi Agricultural University, the city-governed forestry research institute in Jingdezhen, the orchards in Tiantou forest administration station, the Laicun Town of Ningdu County, and the orchard in Tianjingyuan village of Nancheng County were taken as test sites, and grain amaranth was planted in the vacant spaces of forestland and orchards. After it had grown, grain amaranth was cut to feed livestock and poultry, achieving good ecological and social benefits. First, the benefits were manifested in the modification of the micrometeorology. Table 11.19 illustrates the increased surface vegetation coverage, which forms a multi-layer structure, the reduced direct solar radiation to the ground surface, and

Table 11.19 Changes in micrometeorology in grain amaranth-interplanted forestland

Observation area	Temperature (°C)		Air humidity (%)		Ground temperature (°C)					Highest surface temperature (°C)	Lowest surface temperature (°C)
	50 cm	100 cm	50 cm	100 cm	0 cm	5 cm	10 cm	15 cm	20 cm		
Red maple + grain amaranth	32.5	32.1	73.3	67.0	33.3	30.4	30.3	30.1	29.4	51.5	24.8
Pittosporum + grain amaranth	31.3	31.6	71.0	65.3	35.6	31.5	31.0	30.7	29.9	49.5	23.0
Contrast area of *Pittosporum*	31.7	31.6	69.7	68.3	35.6	32.4	32.2	31.1	30.8	51.4	24.0

the decreased maximum and average temperature of the ground surface. The luxuriant branches and leaves of grain amaranth helped to increase air humidity under the canopy; the air humidity at 50 cm aboveground was significantly higher than that in the control area.

11.4.3 Improving the Effect of Composite Management System of Tea–Forest on Micrometeorology

Tea produced in tea-producing areas in low mountainous and hilly regions of the middle and lower reaches of Yangtze River often has poor quantity due to strong sunlight and low humidity. Generally speaking, the light saturation point of tea plants is 50,000 lx; when it exceeds the point, leaf photosynthetic intensity of tea plant will decrease with the increase of light intensity. The optimum temperature for tea growth is 20–30°C; in the range of 10–25°C, the growth of shoot increases as temperature increases. When it exceeds 30°C, the growth will be suppressed. The proper air humidity for tea plants is 80%. Air humidity should be kept above 70% within 20 days before tea plucking. When air humidity is below 60%, the respiration rate of tea plants will increase, resulting in the reduction of tea leaf production and in inferior quality leaves. To establish high-yield and high-quality tea gardens in the above tea-producing areas, a feasible and effective method might be to interplant tree species in the tea garden to provide moderate shade, creating appropriate ecological factors, such as light, temperature, and humidity, in addition to choosing the appropriate ecological environment.

Fifteen test points were randomly selected in the tea garden for use of the interplanting model of tea and persimmon in Tongdan village, Jing County of Anhui Province; another test point was selected in the center of a pure tea garden. The micrometeorological factors of the tea garden were simultaneously measured at these test points with the height of 1.2 m (near the tea canopy). The tests were conducted every 2 h from 06:00 to 18:00 for three successive sunny days. The results showed that, compared to the pure tea garden, the light intensity of intercropped tea garden clearly decreased. The decreases in spring and summer tea times were 9.5 and 39.4%, respectively. At the same time, the component of diffusive light increased in the intercropped garden, which benefited the growth of tea plants and improved the tea quality. In spring tea time, the persimmon tree was at the initial growing stage of leaves with a weak shading effect. Therefore, the temperature of the picking surface in the intercropped garden was just 0.62°C lower than that of the pure garden. Temperature reduction in summer tea time in the intercropped garden was very obvious, with the average daily temperature of the picking surface being 1.95°C lower than that of the pure garden. Meanwhile, the daily temperature variation was lower than that of the pure garden; the duration of the high-temperature period clearly shortened. All these things were helpful to the growth of new shoots of tea plants. The relative humidity of the intercropped garden in spring and summer increased 6.2 and 11.3%, respectively, over that of the pure garden, which benefited the improvement of tea by keeping the leaves tender. In conclusion, the micrometeorological conditions in the intercropped tea garden were improved.

In the forestland of southern Anhui Province, the slash pine tree stand was established in 1966, and then tea trees were interplanted in it in 1971. At the same time, the pure tea garden was established on a contiguous area with similar terrain and soil conditions; also the tea variety, row spacing, and management measures were exactly the same as the intercropped tea garden. The micrometeorological conditions of the system clearly changed after the introduction of composite management of slash pine and tea (Table 11.20). The daily pattern of change in light intensity of the tea garden with composite management and the pattern of change in the pure tea garden were almost the same, but the light intensity of the tea garden with composite management was clearly lower than that of the pure tea garden. The period of light intensity above 50,000 lx was about 2 h for the former, whereas that of the latter was more than 6 h every day.

The changes of temperature in the tea garden depended on the changes of light, and usually the highest temperature occurred 2 h later than highest light intensity. Table 11.20 shows that the daily highest temperature of the composite model of tea and forest was below 30°C in the hilly region of the middle and lower reaches of Yangtze River, whereas there were 2 h a day in the pure tea garden with temperatures above 30°C. Moreover, the variations in the relative humidity in these two types of tea gardens were different. There were more than 4 h during daytime with air relative humidity above 80% in the composite tea garden with tea and pine but only about 2 h for the pure tea garden; the average relative air humidities during daytime of the two gardens were 73 and 64%, respectively. Therefore, with the composite management of tea and pine, light intensity and air temperature in tea garden were decreased, while air humidity was increased, which provided a favorable micro-environment for the growth of tea and effectively improved tea production and quality.

In the tea garden with tea and Chinese tallow (*Sapium sebiferum*) intercropped, micrometeorological factors and soil conditions were quite different from those of the pure tea garden, due to the shading of Chinese tallow, interpenetration of the

Table 11.20 Daily variations of the micrometeorological factors in tea gardens with different management models

	Light intensity (lx)		Temperature (°C)		Air humidity (%)	
Test time	Intercropped garden	Pure garden	Intercropped garden	Pure garden	Intercropped garden	Pure garden
6:00	800	24010	12.4	12.2	95	90
8:00	13200	26700	13.4	14.7	91	77
10:00	46500	8300	18.8	20.0	79	61
12:00	8200	101000	25.4	27.2	67	57
14:00	48900	87200	29.2	31.2	55	50
16:00	34100	50000	26.9	30.4	61	55
18:00	4500	5500	23.8	26.8	65	60

root systems, and the increase of litter and enhancement of litter quality. Climate factors such as light intensity, temperature, ground temperature, and air humidity were measured every 2 h from 6:00 to 18:00 for three successive sunny days, for the study of microclimatic changes of tea garden intercropped with Chinese tallow.

Test results (Fig. 11.2) showed that, compared with that of the pure tea garden, light of the intercropped tea garden clearly decreased, whereas the proportion of diffusive light increased, which benefited the growth of tea plant and the improvement of tea quality. In spring tea time, the light intensity in intercropped tea garden varied within the range of 4.8–85.8 klx. The average daily light intensity declined by 11.6% compared with pure tea garden. It was more obvious in summer tea time, with light intensity decreasing by 40.7% compared with pure tea garden. The measured results of temperature indicated that the average daily temperature of the plucking surface of spring tea was 22.4°C, only 0.8°C lower than that of pure tea garden, and 1.9°C lower in summer tea time. Meanwhile, the high-temperature period in the

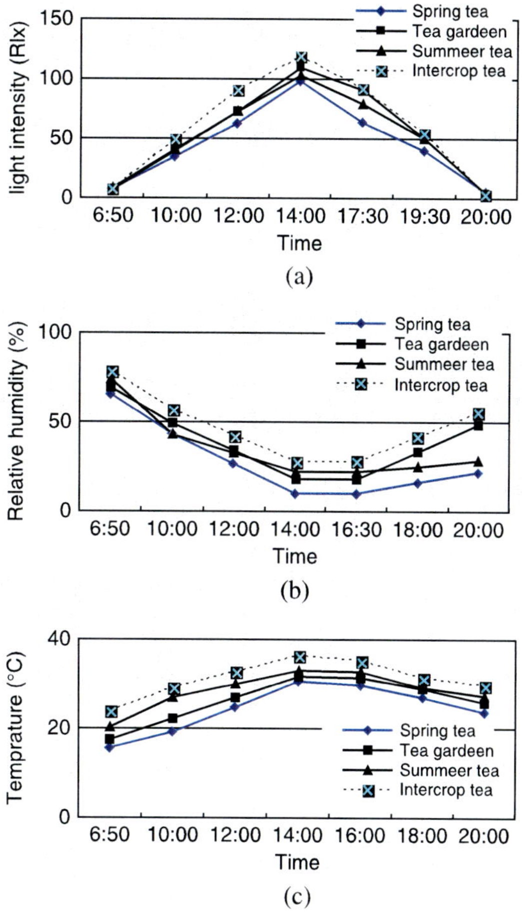

Fig. 11.2 Daily variation of micrometeorological factors in different tea gardens

intercropped tea garden shortened remarkably, which was conducive to the growth of new tea shoots, improving the production and quality. At the same time, as shown by Fig. 11.2, the intercropped tea garden clearly had increased relative humidity, with an increase of about 10% in summer tea time, which also benefited the improvement of tea quality. As for soil conditions, intercropped tea garden and pure tea garden were different in soil temperature at different depths. For example, the temperature of the 0–10-cm surface soil layer in the pure tea garden was clearly higher than that of intercropped tea garden, and the difference declined gradually below 10 cm. In summer tea time, average daily temperature of the 5-cm-deep soil layer in the intercropped tea garden was 23.4°C, or 1.8°C lower than that of the pure tea garden, and the highest daily soil temperature was 27.5°C, which was 3.5°C lower than that of the pure tea garden. The difference of geothermal conditions led to the improvement of growth condition in the intercropped tea garden to some extent.

The results indicate that the intercropping model of tea and forest achieved desirable ecological benefits. Many other intercropping models had similar ecological benefits, such as the stereoscopic management model of tea and Chinese tallow, the intercropping model of persimmon and tea, and the intercropping model of slash pine and tea. In the southern tea-producing areas, especially in low mountainous and hilly regions with flat terrain, aside from selecting the appropriate ecological environment for tea growth as carefully as possible, it will also be an effective method and model for the establishment of high-yield and good-quality tea garden to intercrop tree species, providing moderate shading for tea garden and creating favorable ecological factors of light, temperature, and humidity.

11.4.4 The Effect of Stereoplanting Pattern in Orchards on Micrometeorology

In view of severe soil and water loss and poor soil fertility in new orchards in the low hilly regions of the middle and lower reaches of Yangtze River, a new comprehensive control model of stereoplanting was adopted for the orchards. It included the addition of green manure, grass, and economic crops to the traditional single cultivation manner, which made the structural layout of biological populations inside orchards more reasonable. It increased the vegetation coverage of green plants. It not only generated great output and economic benefits but also brought about greater benefits of ecological protection.

At the four experimental sites in four orchards – the orchard of Tiantou forest administration station, the Youcun orchard in the town of Laicun, the Shuibao orchard in the town of Changsheng, and the navel orange orchard in Liangdu of Ningdu County in the mountainous areas of southern Jiangxi Province – the artificial planting of Bahia grass (which had a biomass of several and even over 10 times higher than that of local wild grass) on slope walls or hillside ditches played an effective role in protecting the slopes. In addition, green manure, such as common cowpea (*Vigna unguiculata*), *Crotalaria* (*Crotalaria pallida*), lobular *Crotalaria* (*Crotalaria striata*), pigeon pea (*Cajanus cajan*), and radish (*Raphanus*

Table 11.21 Changes of environmental factors in the micrometeorology of stereoplanting pattern in Ningdu orchard

Test sites	Temperature (°C)		Relative humidity (%)		Variation amplitude of soil temperature at 5–20 cm (°C)	Highest surface temperature (°C)	Lowest surface temperature (°C)	Field capacity (%)	Average surface temperature (°C)
	50 cm	100 cm	50 cm	100 cm					
Fruit + amaranth	29.86	29.88	79.4	78.2	30.3–28.7	46.86	16.6	24.38	31.31
Fruit + bean	29.45	29.56	81.4	81.4	28.6–27.8	42.86	16.37	21.98	29.72
Tree + green manure (grass)					31.0–29.0	47.0	18.8	2.17	28.94
Control area (without interplanting)	30.0	30.3	78.0	75.0	31.36–29.2	47.86	16.7	18.46	32.34

sativus), was suitable to be planted on granite-eroded inferior lands, and these could quickly cover the ground surface and control soil and water loss, with rapid effects and large benefits. Even under the conditions without irrigation or with severe drought, they could grow normally and achieve satisfying production of fresh leaves. Table 11.21 shows the effect of several different stereoplanting patterns of orchards on environmental improvement.

The results indicate that the stereocomposite management of the orchards effectively decreased the speed of air convection, increased relative air humidity, greatly increased the water-retaining capacity of orchard soil, decreased the average temperature, and reduced the variation in amplitude of soil temperature, all of which were conducive to the growth of plants.

Stereoscopic planting models of orange–economic crops, orange–fertilizer (green manure)–economic crops, and orange–grass–economic crops fully used the large spatial and temporal gaps before the fruit-bearing time for the stereoplanting of beans, peanuts (*A. hypogaea*), sweet potatoes (*Ipomoea batatas*), rapes (*B. campestris*), radish etc., increased vegetation coverage, and formed multi-layer composite populations, which could effectively decrease the ground temperature in dry summers with high temperature, improve water storage and conservation functions of the orange orchard, regulate soil water conditions, and improve the micrometeorology in the orchard, thus benefitting the growth of oranges. At the same time, the straw, litter, root system, and green manure of these crops were directly returned to orange orchards, effectively increasing soil organic matter, especially the symbiotic and nitrogen-fixing functions of rhizobia of legume crops, soil nitrogen content, and improving soil fertility. Through cultivating and weeding of interplanted crops, soil maturity and improvement can be accelerated, leading to remarkable ecological benefits.

From 27 July to 5 August 1996, a 10-day observation of the micrometeorology was conducted at stereoplanting areas of 3-year-old navel oranges (*Citrus sinensis*) + grain amaranth, navel oranges + soybean, and a control area of navel oranges without stereoplanting at the navel orange base of Tiantou orchard in Ningdu County. The observed factors included the highest surface temperature, the lowest temperature, surface temperature, dry and wet bulb temperature at the height of 20, 50, 100 cm above the ground, and ground temperature.

Table 11.22 Differences in the micrometeorological factors between stereoplanting and control area (CK)

Test plot	Surface temperature (°C)	Highest temperature (°C)	Humidity at the height of 100 cm %	Temperature at the height of 100 cm (°C)
Orange + amaranth	−1.6–+1.8	−0.2–+2.8	+1–5.6	−0.1–+1.7
Orange + bean	−4–+1.8	−3–6.9	+1–8.7	−0.1–0.7

Table 11.22 shows that, compared with the control area, the surface temperature of the navel orange + bean area decreased to different extent with the maximum reduction at 4°C, while the highest ground temperature of navel orange + bean area decreased by 3–6.9°C. As for relative air humidity at different heights aboveground, this was increased to different extents in both the navel orange + grain amaranth area and navel orange + bean area compared with control area, with a remarkable increase in height of 100 cm. The daily average temperature at different heights in stereoplanting areas declined to different degrees compared with control area, with an obvious decline in the navel orange + bean area at the height of 100 cm. The reasonable stereoplanting could effectively improve the micrometeorology of the orange orchard, benefiting the growth of oranges.

Reference

Zhang HJ, Kitahara H, Endo T (1994) The effect of the several kind of litters to the roughness coefficient. J Water Soil Conserv 8(4):4–7

Chapter 12
A Study on Plant Roots and Soil Anti-scourability in the Shangshe Catchment, Dabie Mountains of Anhui Province, China

Abstract The distribution of root biomass was studied in different layers (0–10, 10–20, 20–30, and 30–40 cm) of soil depth with the plot method in different types of plants in the Shangshe catchments in the Dabie Mountains of Anhui Province. Samples of root system were taken to the lab to measure their quantity and length. In each plot, soil anti-scourability in corresponding soil layer was measured with a C.C. Suoboliev anti-scourability instrument. The results showed the following. (1) The root system was mainly distributed in the 0–40-cm layer of soil, and the root quantity was the greatest in the surface layer of soil. The fine roots smaller than 1 mm in diameter were absolutely predominant in root length. (2) In the same section, the values of the anti-scourability index of surface layer were larger than those of other layers in different types of plants. The tree root system, especially the fine roots less than 1 mm in diameter, had powerful function on soil loss control. The correlated coefficients between the length, the quantity, and the density of fine roots and the anti-scourability index were 0.817, 0.716, and 0.643, respectively. The mean length of fine roots was a key factor to soil anti-scourability index. The anti-scourability index was closely correlated with non-capillary porosity of each different type of soil, indicating that forest had a soil stability function because their root system improved soil physical properties and finally created a biosoil system with anti-scourability.

12.1 Introduction

The special topography and the geomorphology of mountains provide specific boundary conditions and dynamic sources for the confluence of rainfall, while the accompanying slope erosion, gully erosion, and debris flow gravity erosion inevitably endanger the mountainous area's agriculture, water conservation capabilities, and the safety of people's lives and property (Wang and Xie, 1998). Ecological environmental deterioration caused by soil and water loss inevitably affects the economic and social sustainable development of mountainous and related areas. It has commonly been acknowledged that "heavy rain or torrential rain in the whole catchment, low ability in flood control and discharge in middle and lower reaches, vegetation destruction in the middle and upstream areas" (Zhang et al., 1999; Zhan et al., 2000; Wu, 1999; Cheng, 1999) were the main reasons for the catastrophic flooding in the whole Yangtze River basin in 1998.

J.C. Zhang et al., *Theory and Practice of Soil Loss Control in Eastern China*,

DOI 10.1007/978-1-4419-9679-4_12,

The Dabie Mountains are located on the border of E (Hubei Province), Yu (Henan Province), and Wan (Anhui Province), with an area of 13,800 km^2, 46% belonging to the Yangtze River basin and 54% belonging to the Huai River basin (Yu and Shi, 1991). The Yangtze River widens in the middle and lower reaches with great decrease in the elevation gradient, which promotes the deposition of sediment. If flooding occurs in the whole basin, the low ability for flood discharge below the middle reach can result in increased flooding in the upper reach and "causes successive increases of flood–water level of subsequent floods to much higher levels than previous floods under the similar flow" (Li and Ni, 1998), and severely aggravates the persistence and harmfulness of flooding. Therefore, it is of great significance to conduct a study on soil and water conservation and the hydrological effects of plants in different types of lands in the Dabie Mountains.

Plant roots are the main plant components having a function in the symbiosis of the soil–vegetation ecosystem (Zhou and David, 1998). They function in the improvement in the soil environment in two main ways. First, plant roots undergo extensive material exchange with soil through rapid turnover of large amounts of fine roots, contributing to the soil self-fertilization function. Second, plant roots stabilize sand and soil by means of a fine network formed by interpenetration of roots of different diameters. This theory has been fully validated by the following work: Zhu's (1982) study on soil permeability and anti-scourability in the Loess Plateau, the studies of Li et al. (1993) and Zhang et al. (1994) on the relationship between soil anti-scourability and plant roots, and the study of Jiang et al. (1979) on factors influencing soil scourability. Up to now, however, there has not been any systematic report on the relationship between different types of plant roots, or the same plant's roots at different ages, and soil anti-scourability at the catchment scale. Anti-scourability of the root systems of different types of plants has yet to be verified by investigation on soil loss in runoff fields, at the sub-catchment, or, finally, the catchment scales for different types of plants. In order to perform a systematic study, beginning in 1999 the authors have set up soil and water loss observation sites at four scales: the catchment, sub-catchment, plot, and micro-plot scales in different types of lands in the Shangshe catchment, Yuexi County of Anhui Province (Zhuang et al., 2004), Meanwhile, studies have been conducted on the relationship both between plant roots in different types of lands, and between roots at different ages on the same plant, and soil anti-scourability and soil permeability. This chapter is a report on part of that study. Based on the analysis of correlations between root features at different ages of different types of plants and both soil physical properties and the soil stability function, this study aims to reveal the internal mechanism of the soil stability function by different types of plant roots in the Dabie Mountains, to make a comprehensive judgment on soil stability function of different types of plants, and to provide a basis for the choice of plant types for the Changjiang conservation forest in the Dabie mountains.

12.2 Research Methods

12.2.1 Choice of Different Types of Plants and an Investigation on Soil and Roots

In the Shangshe catchment, well-managed, middle-aged masson pine forest (*Pinus massoniana*) (A, 18 years old), mature masson pine forest (B, 41 years old), middle-aged Chinese fir forest (*Cunninghamia lanceolata*) (C, 10 years old), mature Chinese fir forest (D, 18 years old), and bamboo forest (*Phyllostachys edulis*) (E) with canopy density greater than 0.8 were chosen for study. Study plots of 20 m × 20 m (except for C, which was 15 m × 20 m, and E, which was 10 m × 20 m) were set up for enumeration surveys. The basic conditions of the study plots are listed in Table 12.1.

Table 12.1 Basic conditions of five forest types

Plant type	Soil depth (cm)	Average height of tree (m)	Average diameter of tree at 1.3 m (cm)	Density (N hm^{-2})
A	43	6.2	10.3	1250
B	62	13.5	23.8	750
C	45	8.3	12.1	1725
D	85	18.2	21.9	1325
E	85	16.6	9.2	1956

In each study plot, 1-m-wide soil profiles with roots were dug, recorded with conventional soil investigation methods, and observed in different layers (0–10, 10–20, 20–30, and 30–40 cm) for extraction of soil samples. Indices, such as soil density, in different soil layers were analyzed in the lab. Meanwhile, soil samples of 25 cm × 25 cm × 10 cm were dug to different layers (0–10, 10–20, 20–30, and 30–40 cm) with replicates. Roots were picked out and cleaned. Average dry biomass and length of plant roots in different layers were measured in the lab.

12.2.2 Measurement of Soil Anti-scourability

Soil anti-scourability was measured with a C.C. Suoboliev anti-scourability instrument. At 1 kg pressure, using a water column of 0.7 mm diameter to strike a soil layer for 1 min until water erosion points occurred, the reciprocal of the mean value of every 10 soil erosion points' depth × width was defined as the anti-scourability index of the soil layer (Zhang and Hu, 1996). In forests, because of the influence of the litter and humus layers, soil anti-scourability was potentially unlimited (Jiang, 1979; Ding et al., 2002), so the litter layer was removed from the surface soil layer. Soil anti-scourability was measured in different layers (0–10, 10–20, 20–30, and 30–40 cm) and the plant root density in different layers was measured as well.

12.3 Results and Analysis

12.3.1 Distribution Characteristics of the Root Profiles of Different Types of Plants

Vertical distribution characteristics of the lengths and quantities of roots of different types of plants are listed in Fig. 12.1.

Figure 12.1 shows that the root system was mainly distributed in the 0–40-cm soil layer and that the lengths and quantities of roots less than 1 mm in diameter were greater in the surface soil layer than in the other soil layers. By calculation, total root lengths of different types of plant in the surface soil layer (0–10 cm) followed the following order: mature Chinese fir forest > bamboo forest > middle-aged Chinese fir forest > mature masson pine forest > middle-aged masson pine forest. Length ratios of roots of different diameters (< 0.5 mm:0.5–1.0 mm:1.0–2.0 mm: 2.0–5.0 mm) were as follows: mature Chinese fir forest, 1:0.4:0.01:0.004; middle-aged Chinese fir forest; 1:0.2:0.1:0.04; bamboo forest, 1:2.8:0.8:0.1; mature masson pine forest, 1:0.4:0.2:0.2; and middle-aged masson pine forest, 1:0.3:0.1:0.04. For all the forests except bamboo forest, roots less than 0.5 mm in diameter were absolutely predominant in total root length in the 0–10-cm soil layer; while for bamboo forest, roots within 0.5–1.0 mm in diameter were predominant in root length, and their root lengths were 1.8 times of those of roots less than 0.5 mm in diameter. Generally speaking, in the 0–40-cm soil layer, although roots less than 0.5 mm and within 0.5–1.0 mm in diameter contributed weakly to total forest biomass, with the highest proportion in total root biomass being 69.19%, they occupied a great proportion in root length, with the highest proportion being 97.33% and the lowest being 74.26%. The longer the roots are, the wider their range of extension is, crisscrossing to form a network in the soil.

Research results showed that the lengths and quantities of roots less than 1.0 mm in diameter in the 0–40-soil layer of mature Chinese fir forest were much greater than those of the middle-aged Chinese fir forest, which was also reflected to a certain degree in comparisons of mature masson pine forest, middle-aged masson pine forest, and middle-aged Chinese fir forest. But due to the huge spatial variability of the mountainous area, this phenomenon needs to be verified by more empirical data. In view of the twofold influence of fir forest on soil fertility, to improve production, the fast-growing and high-yield functions of fir trees should be fully utilized, because the great length and quantity of their roots less than 1.0 mm in diameter in the surface soil layer can improve soil properties. At the same time, mixed forest should be created to overcome the defect of soil deterioration caused by a pure fir forest (Wu et al., 1994).

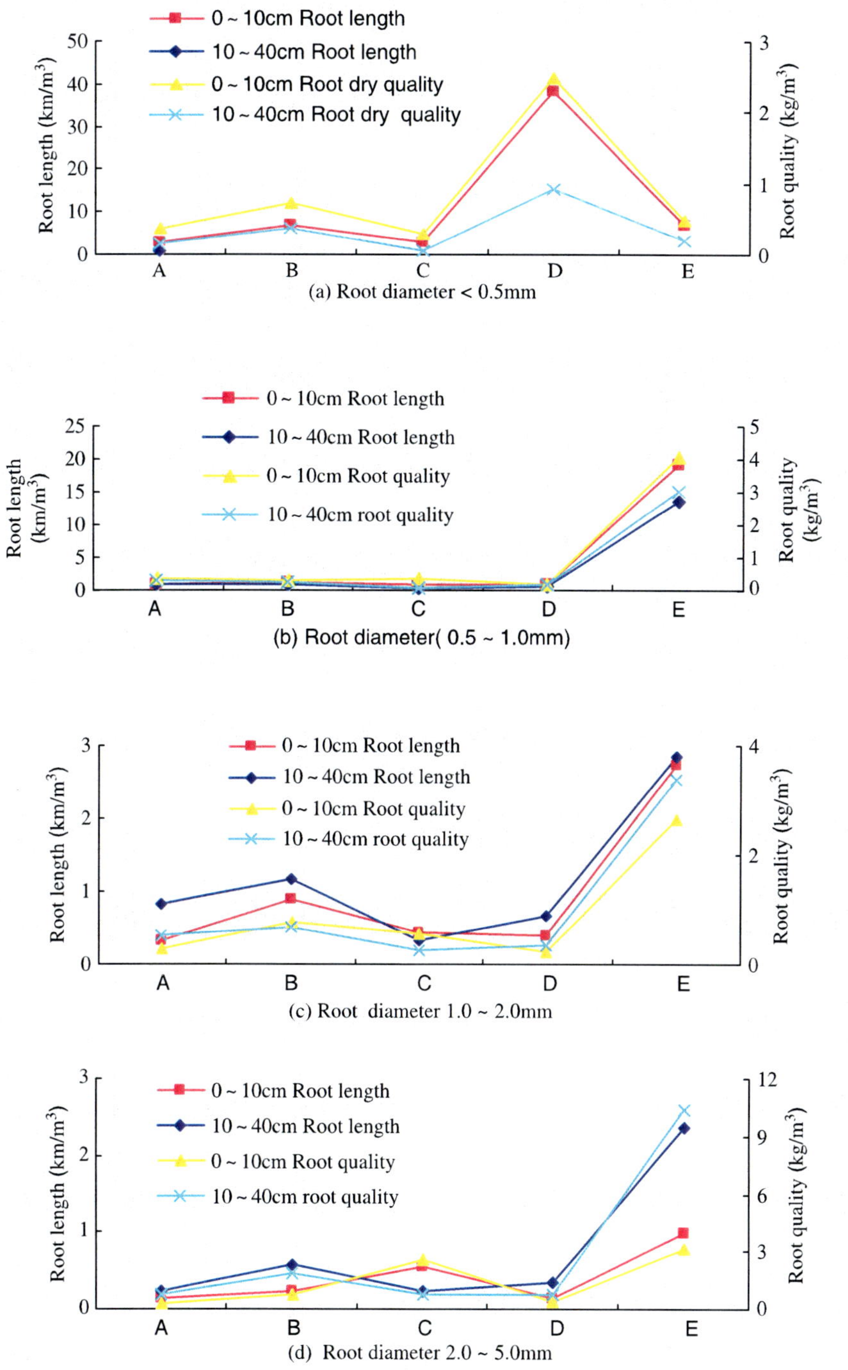

Fig. 12.1 Vertical distribution of lengths and quantities of roots in five diameter classes less than 5.0 mm diameter and for five plant types

12.3.2 Plant Roots and Soil Anti-scourability

12.3.2.1 Soil Anti-scourability Variation in Different Types of Plants and in Different Soil Layers of the Same Plant

Soil anti-scourability refers to the ability of soil to resist mechanical damage caused by agents like runoff and wind. It is affected by soil physical and chemical properties, soil profile structure, and biological factors (Zhang and Hu, 1996). A large number of studies (e.g., Jiang, 1979; Zhu, 1982; Ding et al., 2002) have demonstrated that plant roots, especially roots less than 1.0–2.0 mm in diameter, have the biodynamic function of stabilizing soil structure, increasing water-stable aggregates (> 2 mm) and organic matter content, and creating a soil system with anti-scourability.

The measured values of anti-scourability in the main types of plants in different soil layers in the studied area are listed in Fig. 12.2. The figure shows that soil anti-scourability varied greatly in different types of plants and in different soil layers of the same plant, and exhibited some rules. For the same plant, soil anti-scourability in the surface soil layer (0–10 cm) was greater than that in the subsoil layer, with a difference of 0.1842–0.6678. The difference between the upper and lower layers of the middle-aged Chinese fir forest was the greatest, while that of the mature Chinese fir forest was the lowest. This was related to root quantity in the surface soil layer, especially the large quantity of fine roots less than 1.0 mm in diameter, which have powerful tensile capacity and elasticity. Together with other effects, like the root network, root soil bonding, and root biochemistry, this strengthens anti-scourability (Zhu, 1982; Li et al., 1993; Zhang et al., 1994). Jiang proposed three general types of vertical distribution of soil anti-scourability: (1) the agricultural cultivation type, which has strong anti-scourability in the crop root layer and weak anti-scourability in the surface layer and the layer below the roots; (2) the soil-forming processes type, which has strong anti-scourability in the clay layer and weak anti-scourability in the upper and lower layers; and (3) the biological growth-influencing type, which

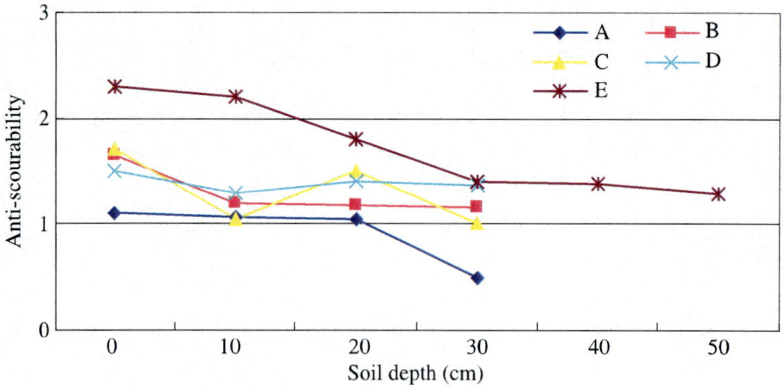

Fig. 12.2 Vertical distribution of anti-scourability in five types of forests

has strong anti-scourability in the surface soil layer and which gradually weakens below the surface layer. Figure 12.2 suggests that soil anti-scourability in different types of plants in Shangshe catchment fluctuated to a certain degree (e.g., soil anti-scourability for mature Chinese fir forest in the 10–20-cm layer (1.0458) was lower than that in the 20–30-cm layer (1.5084)) but was basically in accordance with the "biological growth-influencing type," in which soil anti-scourability in the surface soil layer was the strongest and which gradually weakened below the surface layer.

12.3.2.2 Analysis of the Correlation Between Soil Anti-scourability and Plant Roots

Based on multiple-linear stepwise regression analysis of the soil anti-scourability index (Y) in different soil layers, and the lengths, quantities, and densities of roots of different diameters in corresponding layers, significant linear regression relations have been found between the lengths, quantities, and densities of roots less than 1.0 mm in diameter and the soil anti-scourability index:

$$
\begin{aligned}
Y &= 1.0872823 + 0.13749183X_1 + 0.51438240X_2 \quad && R = 0.8173, \quad && N = 22 \\
Y &= 1.15170742 + 0.00232108X_3 && R = 0.7159, && N = 22 \\
Y &= 1.1200 + 0.0000466X_4 && R = 0.6434, && N = 22
\end{aligned}
$$

where X_1 and X_2 represent the lengths of roots less than 0.5 mm and within 0.5–1.0 mm in diameter, respectively (cm); X_3 represents the quantity of root less than 1.0 mm in diameter (g); and X_4 represents the density of root less than 1.0 mm in diameter (root m^{-2}).

When a significance test was performed on the three regression models, it showed that their regression significance was 0.9999, 0.9998, and 0.9988, respectively, with factor regression significance better than 0.99.

From the above analysis, it can be concluded that the influence of plant roots on soil anti-scourability is mainly displayed as the consolidation, networking, protection, and blocking effects on soil by massive fine roots and root hairs. This had two causes. First, root quantity per unit volume was a function of root diameter. Roots of large diameter were low in density, with small contact area with the soil, so they had low effectiveness on both soil absorption and soil stability, and did not necessarily play a decisive role in improving soil anti-scourability. Second, tough fibrous roots were small in diameter, with a high density and large root length, and they played an incomparably greater role than did coarse roots in soil improvement and the creation of soil system with anti-scourability (Li et al., 1993; Zhang et al., 1994). Since aboveground biomass was related to belowground biomass for the same plant, soil anti-scourability prediction by the above regression equations holds great significance for predicting the soil erodibility factor of K in the USLE (Wischmeier and Smith, 1965; Ruan and Wu, 2002; Zhang et al., 2008).

12.3.3 Analysis of the Correlation Between Soil Anti-scourability Enhancement Value and Plant Roots

The enhancement effect of plant roots on soil anti-scourability actually refers to plant root intensification of soil anti-scourability after eliminating the effect of anti-scourability of a soil layer without a root system. Zeng (1995) proposed in the report "Study on Plant Roots Intensification to Soil Anti-Scourability in Purplish Soil of Guizhou" that the anti-scourability enhancement value of a soil layer with a root system was 84.55% higher than that of the soil layer without the root system. With multiple stepwise regression analysis of the lengths, quantities, and densities of roots of different diameters in the study plot and the soil anti-scourability enhancement value (Y_1), the following regression relationship can be derived:

$$Y_1 = 0.78172823 + 0.13749183X_1 + 0.5138240X, \quad R = 0.8713, \quad N = 22$$

where X_1 and X_2 represent the lengths of roots less than 0.5 mm and within 0.5–1.0 mm in diameter, respectively.

This shows that roots less than 0.5 mm and within 0.5–1.0 mm in diameter played a dominant role in intensifying soil anti-scourability, because they have a much higher density than that of roots more than 1.0 mm in diameter. Reflected in root length, massive numbers of fine roots can fully demonstrate their mechanical networking effects on soil.

12.3.4 Comprehensive Analysis of Soil Stability Function of Root System

A correlation coefficient matrix was constructed of 12 indices – root length (X_1–X_4), biomass (X_5–X_8), soil bulk density in correspondent layer (X_9), capillary porosity (X_{10}), non-capillary porosity (X_{11}), and soil anti-scourability index (X_{12}) of roots in different diameters (< 0.5, 0.5–1.0, 1.0–2.0, 2.0–5.0 mm) in the surface soil layer (0–20 cm) for different types of plants. The results are listed in Table 12.2. The results show that the distribution of the lengths of roots of different diameters was correlated in the 0–20-cm soil layer. The correlation coefficients between the mean length of roots within the 0.5–1.0 mm diameter range and that of roots within the 1.0–2.0 and 2.0–5.0 mm diameter ranges were 0.96 and 0.88, respectively; the correlation coefficient between the mean length of roots within the 1–2 mm diameter range and that of roots within the 2.0–5.0 mm diameter range reached 0.95; this implied that the distribution of roots in different types of plants was quite similar within the 0.5–5.0 mm diameter range. Roots less than 0.5 mm in diameter had a low correlation with other roots in root length and biomass, which suggested that the growth and the distribution of finest root in different types of plants were relatively independent and specific.

Table 12.2 Analysis of correlation matrix

	X_1	X_2	X_3	X_4	X_5	X_6	X_7	X_8	X_9	X_{10}	X_{11}	X_{12}
X_1	1.00	−0.07	−0.05	−0.11	0.99	−0.08	−0.11	−0.14	−0.56	0.34	0.07	0.40
X_2	−0.07	1.00	0.96	0.88	−0.15	0.99	0.95	0.85	−0.25	0.38	0.26	0.39
X_3	−0.05	0.96	1.00	0.95	−0.12	0.93	0.98	0.92	−0.25	0.46	0.11	0.37
X_4	−0.11	0.88	0.95	1.00	−0.19	0.82	0.98	0.99	−0.06	0.35	−0.03	0.20
X_5	0.99	−0.15	−0.12	−0.19	1.00	−0.15	−0.19	−0.21	−0.55	0.37	0.03	0.43
X_6	−0.08	0.99	0.93	0.82	−0.15	1.00	0.90	0.78	−0.25	0.42	0.27	0.46
X_7	−0.11	0.95	0.98	0.98	−0.19	0.90	1.00	0.96	−0.17	0.37	0.11	0.26
X_8	−0.14	0.85	0.92	0.99	−0.21	0.78	0.96	1.00	−0.02	0.29	−0.03	0.13
X_9	−0.56	−0.25	−0.25	−0.06	−0.55	−0.25	−0.17	−0.02	1.00	−0.13	−0.67	−0.38
X_{10}	0.34	0.38	0.46	0.35	0.37	0.42	0.37	0.29	−0.13	1.00	−0.40	0.79
X_{11}	0.07	0.26	0.11	−0.03	0.03	0.27	0.11	−0.03	−0.67	−0.40	1.00	−0.02
X_{12}	0.40	0.39	0.37	0.20	0.43	0.46	0.26	0.13	−0.38	0.79	−0.02	1.00

Soil bulk density was negatively related with the lengths and biomasses of roots in different diameter ranges; the absolute value of the correlation coefficient decreased with the increase of the diameter value. Both capillary porosity and non-capillary porosity were positively related to the lengths and biomasses of roots in different diameter ranges, but with a small correlation coefficient, which implies that different types of plants have their own special ways to improve soil physical properties. The soil anti-scourability index was positively related to the lengths and biomasses of roots in different diameter ranges, with a higher correlation coefficient of about 0.4 for roots within the 0.5–1.0 and 1.0–2.0 mm diameter ranges and a correlation coefficient at 0.2–0.3 for other roots. The anti-scourability index had a significant positive correlation with non-capillary porosity, with a correlation coefficient at 0.93, which convincingly shows that different types of plants shared the similarity of improving soil structure and enhancing soil anti-scourability.

Table 12.3 lists the result of principal component analysis of 12 indexes in the surface soil layer (0–20 cm) for different types of plants. It shows that the first principal component contributed 48%, the second 26%, and the third 16% so that the first three together are 90%. In the first principal component, X_2, X_3, X_4, X_6, X_7, and X_8 are all large positive numbers; therefore, the first principal component reflects the comprehensive characteristics of the length and biomass of roots within the 0.5–5.0 mm diameter range in different types of soil. In the second principal component, X_1, X_5, and X_{12} are big positive numbers, while X_2, X_3, X_4, X_6, X_7, X_8, and X_9 are negative numbers; therefore, the second principal component reflects the comparison of the characteristic value and the soil anti-scourability index for roots less than 0.5 mm in diameter with the lengths, biomasses, and soil bulk density of roots within the 0.5–5.0 mm diameter range in different types of soil. In the third principal component, X_9 and X_{11} are the largest in value, but opposite in sign; therefore the third principal component reflects the comparison between soil non-capillary porosity and soil bulk density.

12.4 Conclusions

(1) In the same soil section, the values of the anti-scourability index of the surface soil layer were larger than those of other soil layers in different types of plants.
(2) Soil anti-scourability was closely related to the tree root system, especially the lengths, quantities, and densities of roots less than 1 mm in diameter, with the regression model's multiple correlation coefficients at 0.8173, 0.7140, and 0.6434, respectively.
(3) In the 0–20-cm soil layer for different types of plants, the correlation coefficients between the mean length of roots within the 0.5–1.0 mm diameter range and those of roots within the 1.0–2.0 and the 2.0–5.0 mm diameter ranges were 0.96 and 0.88, respectively. The correlation coefficient between the lengths of roots within the 1.0–2.0 mm diameter range and that of roots within the 2.0–5.0 mm diameter range reached 0.95. These results showed that

Table 12.3 Principal component analysis

Principal component	Characteristic root	X_1	X_2	X_3	X_4	X_5	X_6	X_7	X_8	X_9	X_{10}	X_{11}	X_{12}	Difference	Ratio	Accumulative contribution rate
1	6.27	0.02	0.38	0.39	0.37	0.00	0.37	0.38	0.35	−0.10	0.23	0.03	0.21	2.90	0.48	0.48
2	3.37	0.47	−0.10	−0.08	−0.14	0.50	−0.08	−0.14	−0.17	−0.26	0.33	−0.06	0.37	1.27	0.26	0.74
3	2.10	−0.18	−0.11	−0.05	0.04	−0.15	−0.11	−0.04	0.04	0.54	0.31	−0.65	0.08	1.28	0.16	0.90

the distribution of roots of different types of plants was quite similar within the 0.5–5.0 mm diameter range. Roots less than 0.5 mm in diameter had a low correlation with other roots in root length and biomass, which suggested that the growth and the distribution of finest root in different types of plant were relatively independent and specific.

(4) Soil anti-scourability in different types of plants was highly correlated with soil non-capillary porosity, which suggested that different types of plants were similar in improving soil structure and creating a soil system with anti-scourability through the root system.

Acknowledgments This study is part of the research project number (30872072) of Chinese National Natural Science Foundation and (2006BAD03A16) of the Chinese national 11th five-year plan of forestry science.

References

Cheng GW (1999) Causes of the 1998 Yangtze flood and strategy for damage control. Scientific strategy and flood in the Yangtze basin, Beijing (in Chinese)

Ding J, Wang ZQ, Chen X et al (2002) Study on increased effect of soil anit-scourability by root system of forest land in Red Soil Hilly Region. J Soil Water Conserv 16(4):9–13

Jiang DS (1979) Study on the anti-scourability of different land use types in the Loess area, Chinese. J Soil Sci 4(2):20–23

Li YT, Ni JR (1998) Influence of sediment transport to water level elevating in the middle reach of Yangtze River. J Basic Sci Eng 6(3):215–221 (in Chinese)

Li Y, Xu XQ, Zhu XM, Tian JY (1993) Plant roots and soil anti-scourability. J Soil Water Conserv 7(3):11–18 (in Chinese)

Ruan FS, Wu XH (2002) A review on the indexes of soil erodibility. Bull Soil Water Conserv 16(4):9–12 (in Chinese)

Wang LX, Xie MS (1998) Hydrological effect of mountainous protection forest and its information system for soil erosion control. Chinese Forestry Press, Beijing (in Chinese)

Wischmeier WH, Smith DD (1965) Predicting rainfall erosion losses from cropland east of the rocky mountains: guide for selection of practices for soil and water conservation planning. In: USDA agriculture handbook, vol 282. US Government Printing Office, Washington, DC

Wu K (1999) The characteristics and the warnings of the Changjiang River in 1998. Prog Geogr 18(1):21–25 (in Chinese)

Wu WD, Liao CH, Liu KS (1994) A study on the soil deterioration of the first generation of artificial fir plantation in Jiangxi Province. Res Soil Water Conserv 1(5):64–69 (in Chinese)

Yu GM, Shi HS (1991) Position and function of the protection forest in the Dabie Mountains of Anhui Province. Thesis of protection forest in the upper and middle reach of the Changjiang River. Chinese Forestry Press, Beijing (in Chinese)

Zeng XB (1995) Study on plant roots intensification to soil anti-scourability in purplish soil of Guizhou. J Mt Agric 14(2):20–24 (in Chinese)

Zhan M, Jia HJ, Wang CC (2000) The flood control function and countermeasures of soil and water conservation. Water Conservancy Sci Technol Econ 6(1):31–33 (in Chinese)

Zhang JC, Hu HB (1996) Water and soil loss in China. Chinese Forestry Press, Peking, China (in Chinese)

Zhang JC, Hu HB, Ruan HH (1999) The present situation and control measures of soil and water loss in Changjiang River Valley. J Nanjing Forestry Univ 23(2):17–22 (in Chinese)

Zhang JC, Kang LX, Lu YS (1994) A study on the root system conserving soil action of tree Belton the seawall in Northern Jiangsu Province. J Soil Water Conserv 8(2):43–47 (in Chinese)

Zhang JC, Zhuang JY, Su JS, Nakamura H, Ishikaw H (2008) Development of GIS-based FUSLE model to predict soil loss in a sub catchment of Chinese fir forest with a focus on the litter in the Dabie Mountains, China. Forest Ecol Manage 255:2782–2789

Zhou Y, David W (1998) Hydrological effect of *Pinus yunnanensis* forest on soil erosion control in the alpine gorge region. J Soil Water Conserv 4(3):31–39 (in Chinese)

Zhu XM (1982) Study on the main types of soil erosion by water and concerning factors. Bull Soil Water Conserv 3(1):40–44

Zhuang JY, Nakamura H, Zhang JC (2004) GIS-based semi-distributed sediment discharge model with a focus on the influence of land use in the Shangshe catchment. J Jpn Soc Erosion Control Eng 57(3):14–22

Chapter 13
Social and Economic Benefits of Forest Reconstruction

Abstract In this chapter, the social and economic benefits brought about by forestry–agriculture composite management, wasteland afforestation, and management of forest sites with good methods are introduced. Planting economic crops, green manure, and green fodder in orange orchards not only improve the growth conditions for oranges but also improves land utilization ratios, increases the output value per unit area, and generates satisfying economic benefits. In particular, the virtuous cycle model integrated planting and breeding by planting grain amaranth (*Amaranthus hypochondriacus*), an excellent fodder rich in protein, in the vacant spaces of the orange orchard, feeding livestock and poultry with its stems and leaves, and then collecting animal manure in a biogas digester. The yield of grain amaranth interplanting in orchards was as high as 45 t ha^{-1}, which could be used to feed pigs and poultry, replacing one-third of the concentrated feed and increasing the quality and smell of meat greatly. The biogas produced by biogas digester with raw materials of grain amaranth straw and manure of livestock and poultry could provide domestic energy and illumination for 20 people of four families living on the orchard farm, reducing environmental pollution and saving huge costs of electricity and fuel. With the increasing income, local residents' awareness of forest reconstruction and environment protection was aroused, and poverty was eliminated.

13.1 Introduction

The studies of restoration and reconstruction of degraded barren ecosystems described here have always been carried out using the principles of paying equal attention to economic and ecological benefits, and making compromises between short-, medium-, and long-term benefits, which sets the restoration process on the path of a virtuous cycle. Vegetation restoration not only affects the forestland itself but also influences the surrounding environment, thus contributing to regional and global ecological balance. With vegetation restoration, the ecological system not only produces remarkable ecological benefits but also generates great economic and social benefits. This chapter describes in detail the social and economic benefits brought about by forestry–agriculture composite management, wasteland afforestation, and management of forest sites with good methods, with the hope that it can arouse people's awareness of forest reconstruction, thus accelerating the restoration of degraded ecosystems and realizing the harmonious relationship between man and nature as rapidly as possible.

J.C. Zhang et al., *Theory and Practice of Soil Loss Control in Eastern China*,
DOI 10.1007/978-1-4419-9679-4_13,

13.2 Economic Benefits of Forestry–Agriculture Composite Management System

There are three primary kinds of ground-level crops in the forestry–agriculture composite management system, namely grain crops, such as soybeans (*Glycine max*), peanut (*Arachis hypogea*), rape (*Brassica campestris*), and corn (*Zea mays*); medicinal species and plants, such as cape jasmine (*Gardenia jasminoide*), Wheel wingnut (*Cyclocarya paliurus*), and *Eucommia* bark (*Eucommia ulmoides*); and forage crops, such as grain amaranth.

13.2.1 Economic Benefits of Forestry–Amaranth–Stockbreeding Composite System

As a forage resource rich in high-quality protein, grain amaranth has many excellent properties, such as high nutritional value, strong stress resistance, fast growth, high yield, strong shoot sprouting ability, good forage quality, and wide application. It is considered the most promising crop of the 21st century. In the forestry–amaranth composite management system, the introduction of grain amaranth resulted in more reasonable material and energy flows and increased biological links and an optimized system structure. The multi-level use of materials was achieved in the system: it fully utilized light energy and nutrients in the forestland (orchard), provided sufficient high-quality forage for stockbreeding, and played a role in replacing cultivation with cropping, improving soil quality, reducing soil loss, improving micrometeorology, and promoting the growth of forest trees (fruit trees). Meanwhile, it could utilize scrap material, reduce contamination, and provide both energy resources and high-quality fertilizer. What is especially important is that grain amaranth can partly replace concentrated feed, thus reducing breeding costs. In addition, with simple planting and low costs, it can increase the organic fertilizer necessary for forestry (fruit industry), both effectively promoting the development of a breeding industry and bringing about the prosperity of forestry (fruit industry). Grain amaranth maximizes biomass production available from the system. After digestion by livestock, the waste can be returned to forestland, increasing the intermediate links of the food chain, enhancing the utilization and conversion rates, and providing more products for the market. When the capital is reinvested in forestry, fruit industry, and agriculture, more benefits can be achieved.

13.2.1.1 The Effects of Grain Amaranth on the Weight Increase of Pigs and Analysis of the Economic Benefits

The results (Table 13.1) of pig raising with grain amaranth show that the pig weight increase of experimental group I (feeding with 1424 kg feed made of basal food and 135 kg fresh grain amaranth) was 1.19, 14.06, and 6.35% higher than that of experimental group II (feeding with 1345 kg feed made of basal food and 237 kg fresh grain amaranth) in the early, late, and whole breeding phases,

Table 13.1 Effects of grain amaranth on the weight increase of pigs

Experimental phase	Groups	Days fed	Number of pigs	Weight at the beginning (kg)	Weight at the end (kg)	Weight increase (kg)	Daily weight increase (g)
Early phase	Control group	57	8	25.78	61.48	35.70	626
	Experimental group I	57	8	25.68	59.54	33.86	594
	Experimental group II	57	8	26.04	59.50	33.46	587
Late phase	Control group	41	8	61.48	88.29	26.81	654
	Experimental group I	41	8	59.54	86.47	26.93	657
	Experimental group II	41	8	59.50	83.13	23.63	576
Whole experiment	Control group	98	8	25.78	88.29	62.51	638
	Experimental group I	98	8	25.68	86.47	60.79	620
	Experimental group II	98	8	26.04	83.13	57.09	583

respectively. Table 13.2 lists the consumption of grain amaranth and economic benefits. Compared with the control group (feeding with 1582 kg feed made of basal food and no fresh grain amaranth), except for the slight weight increase in experimental group I, the other experimental groups all indicated reduced weight increases. This had two causes: (1) Because the straw powder from grain amaranth was made up of straw left after seeds were harvested, its quality was poor, with average crude protein at (12.25 ± 2.72)% (9.56–15.0%) and crude fiber at (23.67 ± 5.78)% (17.0–27.2%). Also, these amounts were unevenly distributed, which was also reflected by the weight increase situation of experimental group I in the early and late phases. (2) Because the basal diet for pigs was made up of early cropping rice with high crude fiber content, and grain amaranth straw powder also had a high crude fiber content (22.67%), the total crude fiber content in the diet was too high (10% of these groups were 8.66 and 9.57% in early and late phases, respectively; 15% of these groups were 9.59 and 10.35% in early and late phases, respectively), affecting the feeding benefits of grain amaranth. If a corn diet had been adopted, which had a relative low crude fiber content (4% in early phase and 5% or so in late phase), the results might have been better.

Table 13.2 shows that, compared with the control group, the concentrated feed consumed for a 1 kg weight increase in experimental groups I and II was reduced by 7.25 and 9.54%, respectively, in the early phase, 10.28 and 3.40%, respectively, in the late phase, and 7.28 and 6.33%, respectively, over the whole experiment. And compared with the control group, the feed cost for a 1 kg weight increase was reduced by 3.29 and 5.43%, respectively, in the early phase. In the late phase, the cost was reduced by 7.90% in group I (10%) but increased by 0.82% in group

Table 13.2 Analysis of feed consumption and economic benefits

Experimental phase	Groups	Total weight (kg)	Feed consumption (kg)	Grass consumption (kg)	Ratio of feed to meat*	Feed price (yuan kg^{-1})	Feed cost (yuan)	Grass cost (yuan)	Weight increase cost (yuan kg^{-1})	Costs (±%)
Early phase	Control group	285.60	747.60		2.62:1	1.69	1263.44		4.42	
	Experimental group I	270.88	672.84	51.96	2.48:1	1.69	1137.10	20.78	4.27	−3.29
	Experimental group II	267.68	635.46	112.14	2.37:1	1.69	1073.93	44.86	4.18	−5.43
Late phase	Control group	214.48	835.00		3.89:1	1.56	1302.60		6.07	
	Experimental group I	215.44	751.50	83.50	3.49:1	1.56	1171.56	33.40	5.59	−7.90
	Experimental group II	189.04	709.75	125.25	3.75:1	1.56	1107.21	50.10	6.12	+0.82
The whole phase	Control group	500.08	1582.60		3.16:1		2571.04		5.13	
	Experimental group I	486.32	1424.34	135.46	2.93:1		2308.66	54.18	4.85	−5.49
	Experimental group II	456.70	1345.21	237.39	2.95:1		2181.14	94.96	4.98	−2.92

*1. Ratio of feed to meat = concentrated feed/weight increase

2. The price of grain amaranth was calculated as 0.4 yuan kg^{-1}

3. ±% of cost referred to the percentage of cost increase (+) or (–) decrease compared with control group

II (15%). Over the whole experiment, the cost was reduced by 5.49 and 2.92%, respectively. Within the experimental groups, the concentrated feed consumed for a 1 kg weight increase in group I was 4.64% more than that of group II in the early phase, 6.93% less than that of group II in the late phase, and 0.68% less over the whole experiment. The feed cost for a 1 kg weight increase in group I was 2.13% more than that of group II in the early phase, 2.12% less than that of group II in the late feed phase, and 2.61% less over the whole experiment.

Compared with the control group, the concentrated feed consumption for a 1 kg weight increase in both experimental groups decreased by different degrees, with a range of 3–10%, except for the late phase of experimental group II. This indicates that feeding grain amaranth can reduce the consumption of concentrated feed, with great significance in developing grain-saving livestock husbandry and solving the problem of feed shortage. When grain amaranth was added to the feed, the cost for 1 kg increase of pig weight decreased 2–8%, generating definite economic benefits. If grain amaranth is used in coordination with proper concentrated feed, it might play a greater role and produce more economic benefits.

13.2.1.2 Influence on Quality of Pigs

Pigs with proper weight and identical shape were selected on 1 January 1999 for the slaughter test, which included two pigs from each group (Table 13.3). The indices of the experimental groups were basically the same as those of the control group, that is, feeding pigs with grain amaranth did not influence pig carcass and pork quality.

13.2.1.3 Comprehensive Economic Benefits of the Forestry–Grain Amaranth–Stockbreeding System

The experimental research showed that each pig needed on average 1.0–2.5 kg grain amaranth each day. With a whole breeding period of 120 days, one pig would eat 240 kg of grain amaranth. Grain amaranth harvested in 667 m^2 (calculated by the yield of the 3000 kg a^{-1} of Ningdu orchard) could feed 12–15 pigs; therefore, 70–100 yuan (US 1\$ = 6.22 Chinese yuan) could be saved per pig and in all 1000–1500 yuan would be saved. At the orchard of Tiantou forestry management station in Ningdu County and the comprehensive rearing farm in Tianjingyuan Town of Nancheng County, grain amaranth was planted between the rows of navel oranges and satsuma mandarins. Then grain amaranth was used to feed pigs and chickens; animal manure was collected in a biogas digester to produce biogas, providing domestic energy and illumination for fruit growers; and biogas slurry was returned to orchards as fertilizer. In this way, a stable, highly efficient ecological orchard was established. The comprehensive benefits of the orchard are listed in Table 13.4.

Because the cyclic system adopted the principle of stereoscopic management, fully utilizing the light space and site conditions of the young forest, it had a higher yield than did other land use methods under the same consumption of resources. At the same time, interplanting grain amaranth played the role of replacing cultivation

Table 13.3 The measured results of pig carcass and pork quality ($n = 2$)

						Half-carcass			
Groups	Live weight before slaughtering (kg)	Empty body weight (kg)	Carcass weight (kg)	Ham percentage	Dressing percentage	Skin proportion	Bone proportion	Fat proportion	Lean meat percentage
Control group	94.1	87.93	69.25	32.3	73.59	8.44	73.59	13.07	15.46
Experimental group I	92.85	86.65	86.65	67.59	30.84	8.1	72.79	12.28	19.49
Experimental group II	89.85	82.7	82.7	66.06	30.25	9.23	73.52	13.01	14.5

Note: dressing percentage = carcass weight/live weight before slaughtering × 100

Table 13.4 Comprehensive benefits of the virtuous cyclic model of fruit–amaranth–pig–biogas

Test site	Yield of grain amaranth (kg hm^{-2})	Number of pigs	Feed quantity per day (kg)	Costs saved (yuan)	Capacity of biogas digester (m^3)	Use of biogas
Tiantou of Ningdu County	4500	20	2	104	20	Domestic energy for 15 people
Tianjingyuan Town of Nancheng County	77,400	200	1–2.5	79	30	Domestic energy for 20 people

with cropping and promoting the growth of forests and fruit trees. Taking the Chinese aralis (*Cornus macrophylla*) forest of Xiushui County interplanted with grain amaranth as an example, ground diameter and tree height were 62.2 and 16.3%, respectively, larger than those of pure forest. Table 13.5 shows the income of stockbreeding in Xiushui County.

The virtuous cycle system contributed to the high yield of forest and stockbreeding development, bringing about remarkable economic benefits. The results of the study on grain amaranth interplanting in orchards and implementation effects of the virtuous cycle system showed that the interplanting yield of grain amaranth could reach 45,000 kg ha^{-1}. It could be used to feed pigs and poultry, replacing one-third of the concentrated feed; 100 yuan could be saved for each pig, thus in all saving 1800 yuan. In addition, the laying rate of pheasants increased by 20%, if fed with grain amaranth. The biogas produced by biogas digester with raw materials of grain amaranth straw and manure of livestock and poultry could provide domestic energy and illumination for 20 people of four families living on the orchard farm. In this way, it reduced environmental pollution and saved huge costs. Including the value of energy, biogas slurry, and biogas residues, it was estimated that the economic benefits generated by the biogas digester in orchards could exceed 1000 yuan in 1 year. Added to the direct economic benefits of meat and poultry produced by stockbreeding, the virtuous cycle remarkably increased the income of orchards, reflecting the advantages of the virtuous cycle of the forestry–agriculture–stockbreeding system.

13.3 Economic Benefits of the Forest–Tea Composite System

Tea plants (*Camellia sinensis*) originated in the remote mountains and dense forests in southwestern China. Due to their long-term evolution, tea plants have formed certain ecological adaptations – they are shade tolerant, like warm weather, need habitats with high air humidity, and require more diffuse light.

The hilly regions of the middle and lower reaches of Yangtze River are the main areas for tea production. But because of the strong light intensity and low humidity in most areas, attempts at improvement of tea quality are difficult; tea quality is

Table 13.5 Income of stockbreeding of the experimental forest in Xiushui County

		Investment				Output		
Species	Number raised	Total (yuan)	Purchasing expenses (yuan)	Concentrated feed (yuan)	Management expenses (yuan)	Gross profit (yuan)	Net profit (yuan)	Investment/output
Goat	30 big goats	2229	1200			4800	2571	1:2.5
	10 small goats					743	743	
Pig	18 adult pigs	8454.5	3805	2560	2089.5	15120	6665.5	1:1.8

Big goats were calculated at the average weight of 20 kg and the market gross price of 8 yuan kg^{-1}; pigs were calculated at the average weight of 120 kg and the market gross price of 7 yuan kg^{-1}

often poor. Practice has proven that it is effective to adopt forest—tea composite management for the establishment of high-yield, good-quality tea gardens in the middle and lower reaches of Yangtze River. This is because, in a tea garden with composite management, in the aboveground part, the trees are tall and need sufficient light, while tea plants are small and shade tolerant. Together, they form a sound spatial structure that is inlaid horizontally and hierarchical vertically. The research results on the underground part showed that the absorptive roots of tea plants are mainly concentrated in the soil layers above 20 cm. This constitutes about 60% of the total roots. The roots of the trees are mainly distributed in the 20–60 cm soil layers. This constitutes about 75–80% of the total roots. In the horizontal direction, absorptive roots of a tea plant are mainly distributed within an area with a diameter of 60–65 cm, while roots of a tree are mainly distributed within an area with a diameter of 130 cm. Therefore, similar to the aboveground part, the underground part of interplanted tea garden also displays the sound structure that is inlaid and hierarchical. In a tea garden with forest–tea composite management, the most remarkable benefits are wood, tea, and other products special to the forest. The products are not only of various kinds but also have wide applications, so they can satisfy many kinds of demands and needs. Moreover, because of the shade of trees, interpenetration of roots, and the increase of litter, both microclimatic factors and soil conditions of composite management system are superior to those of a pure tea garden. At the same time, the relative humidity of interplanted tea gardens is significantly improved, with an increase of about 10% in summer tea time, which benefits the improvement of tea quality.

13.3.1 Economic Benefits of Persimmon (Diospyros kaki)–Tea Composite System

13.3.1.1 Effects of Interplanting Persimmon–Tea on Tea Quality

Compared with a pure tea garden, the light intensity of a tea garden interplanted with persimmon is clearly decreased, with decreases of 10.5 and 39.4% in spring and summer, respectively. Meanwhile, diffuse light in the interplanted tea garden increases, which is conducive to the growth of tea plants and the improvement of tea quality. During the spring tea period, persimmon trees are just beginning to spread leaves with a weak shading effect, so the temperature on the tea picking surface is only 0.62°C lower than that of pure tea garden. During the summer tea period, the temperature decrease is accentuated, with the daily average temperature on the picking surface being 1.95°C lower than that of the pure tea garden; meanwhile, the daily temperature variation is lower than that of the pure tea garden, and the high-temperature period is remarkably shortened, which benefits the growth of tea shoots. The relative humidity of the interplanted tea garden increases by 6.2 and 11.3% in spring and summer, respectively, compared with that of a pure tea garden, which is conducive to maintaining tenderness of tea. Thus the micrometeorology of the interplanted tea garden is improved. The results (Table 13.6) of biochemical

Table 13.6 Tea components and other quality indices in two different tea gardens

Quality index	Tea garden interplanted with persimmon	Tea of pure plantation
Amino acids (%)	2.95	2.23
Soluble sugar	16.40	14.20
Caffeine	3.64	3.51
Flavonoids	9.38	11.21
Catechin	57.39	64.02
Tea extract	46.16	43.12
Appearance color	Jade green	Yellow-green
Liquor color	Jade green and bright	Yellow-green and bright
Aroma	Chestnut aroma lasting longer	Chestnut aroma lasting shorter
Flavor	More mellow	Less mellow
Leaf bottom	Yellow-green, fresh	Yellow-green, less fresh
Total score	89	74

analysis and sensory determination show that the contents of amino acids, soluble sugars, caffeine, etc. were higher than those of a pure plantation, but the contents of compounds that are harmful to tea quality, such as the flavonoids, were lower than those of pure tea garden. And the color, fragrance, smell, and shape of tea in interplanted tea garden were also better than those of pure tea garden. In general, the comprehensive score for tea of an interplanted tea garden was 89, characteristic of excellent tea, while that of pure tea garden was 74, belonging to common tea.

Persimmon–tea composite management can fully utilize the light and soil resources. In addition, with the high biomass per unit area and appropriate micrometeorology, the growth of tea plants and the tea quality are remarkably improved, with a high proportion of excellent tea. Supplemented by the income from persimmons, the yield of the persimmon–tea composite tea garden is remarkably improved compared with that of the pure tea garden. The measured data (Table 13.7) in three successive years show that the income of the composite management tea garden was about double that of the pure tea garden; the investment/output ratio of the former was 1:5.7–6.2 and that of the latter was 1:3.6–4.2.

13.3.1.2 Economic Benefits of Persimmon–Tea Garden

Table 13.7 The output and investment/output ratio in different types of tea gardens (units: yuan hm^{-2})

	Interplanted tea garden					Pure tea garden		
Year	Investment	Output (tea)	Output (persimmon)	Output (total)	Ratio	Investment	Output	Ratio
1993	2250	9525	4470	13,995	1:6.2	1800	7470	1:4.2
1994	2700	10,695	4785	15,480	1:5.7	2025	7580	1:3.7
1995	2925	10,620	6255	16,875	1:5.8	2100	7635	1:3.6

13.3.2 Economic Benefits of Slash Pine (*Pinus elliottii*)–Tea Composite System

13.3.2.1 Influence of Slash Pine–Tea Interplanting on Tea Yield

The results (Table 13.8) of the field experiment lasting for five successive years show that the average tea yield per acre of the slash pine–tea garden with composite management was 6.5% higher than that of the pure tea garden. Tea yield of tea gardens with different degrees of shading (that of pure tea garden was 0) differed significantly from each other. The results (Table 13.9) of multiple comparisons reveal the relationship between tea yield and the degree of shading; the tea yields of tea gardens with different degrees of shading are ranked as follows: shading degree of 30–40% > 20–25% > 0 (pure tea garden) > 45–50%. It can be seen that the shading degree of 30–40% was the most appropriate for a pine–tea composite tea garden; light in tea gardens with heavy shading (a shading degree above 45%) was so weak that it negatively affected the photosynthesis of tea plants. Therefore, tea gardens with forest–tea composite management should be designed with attention to the appropriate density of pine trees.

Table 13.8 Variance analysis of tea yield of tea gardens with different degrees of shading

Source of difference	Degree of freedom	Sum of squares of deviations	Mean square	Mean square ratio	
Among groups	3	119.08	39.49		
Within groups	36	92.93	2.58	14.22	$F_{0.05} = 2.87$
Total	39	212.01			

Table 13.9 Multiple comparison of tea yield

$\overline{x_1}$	$\overline{x_1} - \overline{x_4}$	$\overline{x_1} - \overline{x_3}$	$\overline{x_1} - \overline{x_2}$
$\overline{x_1} = 22.2$			
$\overline{x_2} = 20.2$	4.6^{**}	3.5^{*}	2.0
$\overline{x_3} = 18.7$	2.6^{*}	1.5	
$\overline{x_4} = 17.6$	1.1		

$\overline{x_1}$, $\overline{x_2}$, $\overline{x_3}$, $\overline{x_4}$ represent shade degree of 30–40%, 20–25%, 0, and 45–50%, respectively. Superscripts * and ** refer to 95% and 99% confidence

13.3.2.2 Influence of Slash Pine–Tea Interplanting on Tea Quality

The ecological environment of a tea garden can not only improve tea yield but also greatly affect the contents and ratios of biochemical components in tea and appearance in shape of tea. The results (Table 13.10) of biochemical analysis showed that the contents of amino acid, caffeine, and soluble sugar in tea produced by interplanted tea gardens were higher than those in tea produced by pure tea gardens; the contents of flavonoids and catechin in tea produced by interplanted tea gardens were lower than those in tea produced by pure tea gardens.

Table 13.10 Comparison of tea quality in two kinds of tea garden

	Amino acids (g kg^{-1})	Caffeine (g kg^{-1})	Soluble sugar (g kg^{-1})	Water content (g kg^{-1})	Water extracts (g kg^{-1})	Catechin (g kg^{-1})	Flavonoids (g kg^{-1})
Interplanted tea garden	32.8	36.4	4.7	58.1	442.5	60.53	6.94
Pure tea garden	27.4	34.1	13.1	49.9	421.6	64.72	11.85

13.3.3 Economic Benefits of Tea Gardens with Composite Management

The forest–tea composite systems of both persimmon–tea and slash pine–tea effectively improved tea yield and quality through the improvement of tea habitat, and thereby increased the economic benefits of tea in composite systems. When the value of wood was added, the benefits of forest–tea composite system improved remarkably compared with a pure tea garden, bringing about more economic and socio-ecological benefits to society.

13.4 Economic Benefits of the Forest–Grain Composite System

By fully utilizing the nutritional space and soil fertility conditions, the forestry–agriculture composite system achieved both ecological and economic benefits. Table 13.11 lists the investment/output ratios of several favorable interplanting models in low hilly regions.

The data of Table 13.11 show that the output/investment ratio of all these models was above 2:1, with the largest being the models of *Eucommia* + soybean and Chinese aralis + grain amaranth or Chinese aralis + sesame. Therefore, stereoscopic management generated large economic benefits over the short term, which had a good effect on making up for the long production period of forestry and slow capital turnover rate. It was a good way to combine long- and short-term benefits, realize an increase in value of forestland, increase income of forest farmers, activate their cooperation, and reasonably exploit and utilize resources in mountainous areas.

13.5 Economic Benefits of the Forest–Medicine Composite System

Forest–medicine composite management is an efficient way to increase the benefits of forestland. There can be benefits to planting medicinal plants in a young forest of Chinese fir and making use of different canopy densities to form different kinds of micrometeorology. This involves selecting and planting different medicinal plants

Table 13.11 Analysis of economic benefits of forestry–agriculture interplanting

Interplanting model	Area (hm^2)	Yield (kg)	Unit price (yuan kg^{-1})	Output value (yuan)	Cost (yuan)	Net profit (yuan)	Output/ investment	Remarks
E. ulmoides+soybean	0.67	485.5	3.6	1747.8	582.6	1165.2	3:1	Agricultural products were calculated by the market price in 1994
E. ulmoides+rape	0.68	187.5	10	1875	800	1075.0	2.3:1	
C. macrophylla+maize	0.53	118.5	2.0	236.4	88.65	147.7	2.7:1	
C. macrophylla+grain amaranth	0.53	1850.1	0.1	185.0	76.8	56.8	3.8:1	
C. macrophylla+sesame	0.53	17.5	6.0	105				

Note: The above table did not include the output value of wood to fully utilize the space, improve the utilization ratio of forestland, increase the short-term economic income to support long-term benefits, and resolve such problems as the long production period of forestry, slow return, and heavy investment in cultivation and management of young forests.

To alleviate the limited resource supply, beautyberry (*Callicarpa bodinieri*) has been interplanted in the young forest of Chinese fir since 1996, in line with barren hill afforestation. According to its biological characteristics and local managerial and economic conditions, remarkable economic benefits were achieved, providing a new method of ecological and economic construction of hilly regions.

Tables 13.12 and 13.13 show that planting beautyberry in forests can increase biological production and economic benefits, with the output/investment ratio at 2.23:1–2.75:1. The ratio in 1993 was larger, with 102 yuan more for the net income per hectare. The net income increased from 1878 to 1980 yuan hm^{-2} per year in a short time, achieving the goals of promoting medicinal plants within forest and cultivating forest in combination with medicinal plants, forming the forest–medicine stereoscopic management model that combined short-, medium-, and long-term benefits. In addition, thanks to its strong sprouting ability, beautyberry could be planted once and harvested for many years, with the appropriate harvesting being once a year.

Moreover, yellow gardenia (*Gardenia stenophylla*) and economic forest species of *C. paliurus* and *Eucommia* interplanted in the forest–medicine model could be harvested in the second year. Among these, yellow gardenia has wide use, including use for the extraction of pigments and large demand in traditional Chinese medicine market. The leaves of *C. paliurus* were the major materials for Xiushui "magic tea," always in short supply. *Eucommia* was also a main material for traditional Chinese medicine and the chemical industry, with large market demand. The mixed forest, established in 1994 with 26.67 hm^2 yellow gardenia, 18.67 hm^2 *C. paliurus,* and 1.67 hm^2 *Eucommia*, generated an annual output value of 79,800 yuan of medicine and fresh leaves, with an average profit of 4710 yuan ha^{-1} (Table 13.14). It fully displayed the great production potential of stereoscopic management and comprehensive exploitation.

Table 13.12 Survey of the biomass of beautyberry

Project category	Harvesting position	Quadrat biomass (g)	Total (kg hm^{-2})	Yield (kg hm^{-2} a)	Remarks
C. bodinieri planted in 1992	Aboveground part	1150	128.33	25.53	Quadrat area, 1 m^2
	Underground part	775			
C. bodinieri planted in 1993	Aboveground part	675	80	15	
	Underground part	525			

Table 13.13 Investment/output analysis of beautyberry

	Investment part				Output part				
Project category	Seedling (yuan ha^{-1})	Afforestation (yuan ha^{-1})	Cultivation (yuan ha^{-1})	Total (yuan ha^{-1})	Yield (kg ha^{-1})	Unit price (yuan kg^{-1})	Output value (yuan)	Net income (yuan·ha^{-1}·a)	Output/investment ratio
C. bodinieri planted in 1992	450	150	240	840	5745	0.8	306.4	1878	2.23:1
C. bodinieri planted in 1993	450	150	120	720	3375	0.8	180	1980	2.75:1

Table 13.14 Income analysis of forest medicine and economic forest

Tree species	Area (hm^2)	Leaf and fruit yield of individual plant (kg)	Total yield (kg)	Unit price (yuan kg^{-1})	Economic income (yuan)	Average income per hectare (yuan)	Remarks
Yellow gardenia	26.67	0.01	600	14	6000	630	13.33 hm^2 fruit bearing
E. ulmoides	18.67	0.25	4466.3	16	71,460	9000	8 hm^2 picking up leaves
C. paliurus	1.67	0.20	750	10	7500	4500	Market price in 1994

13.6 Economic Benefits of the Forest–Fruit Composite Management Model

Fruit–green manure–economic crop composite management is a technology combining green manure with the traditional composite management pattern. Its main purposes are to support long-term benefits with short-term income, replace cultivation with cropping, and improve land capability. In particular, through the proper mixing of green manure and economic crops, the purpose is to combine the mutual–complementary and mutual–beneficial relationships in time and spatial sequences among plant populations, to comprehensively consider their main functions of water conservation and soil improvement, to treat and manage by different types, and to achieve vegetation coverage of orchards all year round. This technology applies to newly developed orchards with low fertility and severe soil and water loss.

Planting economic crops, green manure, and green fodder in orange orchards not only improves the growth conditions for oranges, but also improves land utilization ratios, increases the output value per unit area, and generates satisfying economic benefits. For example, in the forest–fruit farm of Tiantou, Ningdu County, 8.33 hm^2 of peanuts, watermelons (*Citrullus lanatus*), etc. were planted in the vacant spaces in a 3-year-old navel orange (*Citrus sinensis*) orchard in 1995, generating an output value of 63,000 yuan by peanuts and watermelons alone, with the net income increased by 31,000 yuan. In 1996, 3.33 hm^2 of watermelons, 3.33 hm^2 of peanuts, 2 hm^2 of grain amaranth, and 0.067 hm^2 of soybeans were planted in an 10-hm^2 orange orchard. The interplanted crops generated an output value of 120,000 yuan, with the net income increased by 50,000 yuan, which implied an average increase of output value of 7500 yuan ha^{-1} and an average increase of net income of 3000 yuan ha^{-1}.

In addition, the virtuous cycle model which integrated planting and breeding was formed by planting grain amaranth, an excellent fodder rich in protein, in the vacant spaces of the orange orchard, feeding livestock and poultry with its stems and leaves, and then collecting animal manure in a biogas digester. In this model, grain amaranth played a "good link" role in the cycle and remarkably increased the economic benefits. In 1996, 22 hm^2 of grain amaranth was planted in the vacant space of the 3-year-old navel orange orchard. In the planting process, phosphorus fertilizer was used for seed dressing, seeding, and transplanting; also, biogas slurry was applied regularly. The average plant height reached 1.4 m in 1 month after transplanting; the individual fresh stems cut for the first time reached 0.9 kg. In all, 90,000 kg fresh stems and leaves were harvested per hectare. By feeding pigs with amaranth stems and leaves for one meal each day, 30% of concentrated feed could be saved, by which, in all, an income of over 1700 yuan could be increased by 17 pigs compared with that when feeding with concentrated feed all the time. Feeding American pheasants with amaranth stems and leaves not only saved concentrated feed but also increased their laying rate by 20%; moreover, their hair color was bright and eggs were of similar size. The biogas produced in a 20-m^3 digester with the manure of pigs and American pheasants, as well as grain amaranth stems and leaves as raw material, could provide domestic and illumination energy for the 15 people of the five families working in the farm; the biogas slurry provided the organic fertilizer for grain amaranth and fruit trees. For the planting of grain amaranth in a navel orange orchard, 300 g of seeds was needed per hectare with a cost of 30 yuan, fertilizers (30,000 kg base fertilizer and 300 kg top dressing) costing 1500 yuan, and management wages (soil preparation, planting, cultivation, and fertilization) costing 1200 yuan; in all, the investment was 2730 yuan. The net income was calculated using the yield of 45,000 kg ha^{-1} of the orchard of Tiantou forestry management station in Ningdu County, with an investment of 0.06 yuan kg^{-1} of fresh grain amaranth, with the price of green fodder at 0.1 yuan kg^{-1}. The calculation showed the net income increased by 3600 yuan ha^{-1}.

13.7 Comprehensive Evaluation of the Economic Benefits of Major Composite Management Models

The studies on restoration and reconstruction of degraded barren ecosystems have always been carried out according to the principles of paying equal attention to economic and ecological benefits, and combining short-, medium-, and long-term benefits, which sets the restoration process on the track of a virtuous cycle. The ecological benefit of vegetation restoration not only affects the forestland itself but also influences the surrounding environment, thus contributing to regional and global ecological balance. With vegetation restoration, the system not only produces remarkable ecological benefits but also generates great economic and social benefits.

Table 13.15 lists the economic benefits of forestry–agriculture–husbandry composite management in the restoration and reconstruction process of four types of degraded barren ecosystems. The results show that the technology of combining treatment and exploitation, and coordinating ecological restoration and economic development can produce considerable economic benefits. Practice has proven that there is strong vitality in the management model of ecological economy for forestry in low barren hills based on principles of ecological economics, and in the forestry–agriculture–husbandry composite ecosystems for different types of degraded barren ecosystems. Xiushui and Yujiang test sites are examples where tree species were reasonably collocated over a 300-hm^2 mountainous area according to biological characteristics, among which the mixed tree–shrub forest was the main provider for long-term ecological and economic benefits. It had a significant role in improving bad habitats of bare rock and couch grass barren hills and treating soil and water loss, and could also produce large quantities of commercial timber, resulting in long-term benefits. The medicinal plants and economic forest planted on hills were short- and medium-term projects, with fruits and leaves as the main products, with a long harvest time, and good economic benefits. The forestry–agriculture composite management of planting economic crops and excellent fodder, grain amaranth, to drive a breeding industry in young forests at the foot of hills was a short-term project. Before crown closure of the stand, it could achieve high economic benefits and promote soil improvement.

Table 13.15 Economic benefits of different forestry–agriculture–husbandry composite management models (1994)

Barren hill type	Test site	Income project	Area (hm^2)	Total output value (yuan)	Annual yield (kg)	Net profit (yuan)	Investment (Yuan)	Output/ investment ratio
Bare rock barren limestone mountain	Xiushui	Interplant	3.0	4149.2	2659.1	2601.2	1548	2.68:1
		Medicinal plants	47.1	84,960.0	5816.3	79,800.0	5160	
		Breeding		30,263.0		17,079.5	13,184	
Gravel-couch grass barren hill	Yujiang	Interplant		7550.0	8250.0	6550.0	1000	7.55:1
		Medicinal plants		800.0	200.0	750.0	50	
		Breeding		20,675.0		17,325.0	3350	6.17:1
		Fruit industry		1482.0	655.0	1282.0	200	7.41:1
Scrub barren hill	Pingxiang	Medicinal plants	3.5	9450.0	11812.5	6930.0	2520	3.75:1
	Yiyang	Mixed	15.3	1680.0	1600.0	1140.0	540	3.11:1
Open shrub barren hill	Le'an	Interplant	4.67	19,900.0	9575.0	14,400.0	5500	3.62:1
	Jingong	Fruit industry	2.67	35,898.0	99,720.0	27,885.0	8013	4.48:1

It is the stereoscopic forestry, with diversified management focusing on forest and a composite agriculture–forestry model, that achieved the multi-layer exploitation of degraded barren ecosystems and integrated long- and short-term benefits. With complex components and reasonable structure, the system realized the optimum collocation of land, capital, and technology. The output/investment ratio of forestry–agriculture composite management reached 2.68–7.55:1 at different test sites, and the average income per unit area greatly exceeded the level under present forestland management. The system of integrated economic and social benefits in the restored ecosystem generated high productivity of stereoscopic forestry and thus maintained stable development of the system and accelerated progressive succession.

Index

J.C. Zhang et al., *Theory and Practice of Soil Loss Control in Eastern China*,
DOI 10.1007/978-1-4419-9679-4, © Springer Science+Business Media, LLC 2011